Tom Migliaccio

Power Generation Technologies

Power Generation Technologies

Paul Breeze

AMSTERDAM • BOSTON • HEIDELBERG • LONDON • NEW YORK • OXFORD
PARIS • SAN DIEGO • SAN FRANCISCO SINGAPORE • SYDNEY • TOKYO

Newnes is an imprint of Elsevier

Newnes
An imprint of Elsevier
Linacre House, Jordan Hill, Oxford OX2 8DP
30 Corporate Drive, Burlington, MA 01803

First published 2005

British Library Cataloguing in Publication Data
A catalogue record for this book is available from the British Library

Library of Congress Cataloguing in Publication Data
A catalogue record for this book is available from the Library of Congress

ISBN 0 7506 6313 8

For information on all Newnes publications visit our web site at www.newnespress.com

Typeset by Charon Tec Pvt. Ltd, Chennai, India
www.charontec.com

Transferred to digital printing in 2006

Contents

List of figures

List of tables

1 Introduction to electricity generation

Electricity defines the modern world. Everything that we think of as modern, from electric lights, through radio and television to home appliances, electronic devices, computers and all the paraphernalia of the information age depends for their operation, for their existence, on electricity.

Today the citizens of developed countries take electricity for granted while those of under developed countries and regions yearn for it. Yet the supply of electricity is both a complex and an expensive business. Increasingly, also, electricity has become a security issue. While people untouched by modernity can still live their lives without electricity, a modern industrial nation deprived of its electricity supply is like a dreadnought without it engines. It becomes helpless.

This book is primarily about the ways of generating electricity. It does not cover the means of transporting electricity and delivering it to those who wish to use it. Nor does it treat, except obliquely, the political issues that attach themselves to electricity supply. What it does attempt, is to provide an explanation of all the myriad ways that man has devised to produce this most elusive of energy forms.

The book is divided into chapters with one chapter devoted to each type of electricity generation. The explanations provided are thorough and technical where necessary but do not resort to overly technical language where this can be avoided. Readers, who are seeking a full analysis of the thermodynamics of the heat engine, or the differential equations for solving the problem of turbine flow, will need to look elsewhere.

The aim of the book is to provide a description of *every* type of power generation in an easily digestible form. There are occasional lacunas; there is no description of magneto-hydrodynamic power generation, for example. This was considered too obscure, even for a comprehensive text of this type. Otherwise, all practical and some still experimental forms are included.

History of the electricity generation industry

The roots of the modern electricity generating industry are to be found in the early and middle years of the nineteenth century and in the work of men such as Benjamin Franklin, Alessandro Volta and Michael Faraday. Faraday, in particular, was able to show the relationship between electricity and magnetism, a relationship that makes it possible to generate electricity with moving machinery rather than take it from chemical batteries as was the case in his day.

The widening understanding of electricity coincided with the development of the steam engine, and the widespread use of gas for fuel and lighting. In the USA, Thomas Edison developed the carbon filament that produced light from electricity. Similar work was carried out in the UK by Sir Joseph Swan.

Lighting offered the first commercial use for electricity, but it was an insufficient foundation for an industry. What accelerated the growth of electricity generation was its use for traction power. Electric trams for urban transport and the underground railway system in London were the kinds of projects that stimulated the construction of large power stations in the last two decades of the nineteenth century.

Its origins may be in the nineteenth century, but few would dispute that the growth of the electricity industry was a twentieth century phenomenon. There is little doubt, too, that it will become the world's most important source of energy. Vital modern developments such as computers and communications are impossible without it. It is worth remembering, however, that most of the key elements necessary for electricity generation, transmission and distribution were developed during the century before last.

The evolution of electricity generation technologies

The earliest power stations used reciprocating steam engines to generate power. These engines were not ideal for the purpose because they could not easily develop the high rotational speeds needed to drive a generator effectively. This difficulty was eventually overcome with the invention of the steam turbine by Sir Charles Parsons in 1884. Fuel for these plants was usually coal, used to raise steam in a boiler.

Hydropower also entered the power generation mix at an early stage in the development of the industry. Much of the key work on different turbine types used to capture power from flowing water was carried out in the second half of the nineteenth century.

By the beginning of the twentieth century both the spark-ignition engine and the diesel engine had been developed. These too could be used for making electricity. And before World War II work also began on the use of wind turbines as a way of generating power. But until the beginning of the 1950s, steam turbine power stations burning coal, and sometimes oil or gas, together with hydropower stations, provided the bulk of the global power generation capacity.

In the 1950s the age of nuclear power was born. Once the principles were established, construction of nuclear power stations accelerated. Here, it was widely believed, was a modern source of energy for the modern age; it was cheap, clean and technically exciting.

Nuclear power continued to expand rapidly in the USA up to the late 1970s. In other parts of the world, uptake was less rapid but Great Britain, France and Germany invested heavily. In the Far East, Japan, Taiwan and South Korea worked more slowly. Russia developed its own plants and India began a nuclear programme, as did China.

From the end of the 1970s the once lustrous nuclear industry began to tarnish. Since then its progress has slowed dramatically, particularly in the west. In Asia, however, the dream remains alive.

At the beginning of the same decade, in 1973 to be precise, the Arab– Israeli war caused a major upheaval in world oil prices. These rose dramatically. By then oil had also become a major fuel for power stations. Countries that were burning it extensively began to seek new ways of generating electricity and interest in renewable energy sources began to take off.

The stimulus of rising oil prices led to the investigation of a wide variety of different alternative energy technologies such as wave power, hot-rock geothermal power and the use of ethanol derived from crops instead of petrol or oil. However the main winners were solar power and wind power.

Development took a long time but by the end of the century both solar and wind technologies had reached the stage where they were both technically and economically viable. There was considerable reason to hope that both would be able to contribute significantly to the electricity generation mix in the twenty-first century.

One further legacy of the early 1970s that began to be felt in the electricity industry during the 1980s was a widespread concern for the environment. This forced the industry to implement wide-ranging measures to reduce environmental emissions from fossil-fuel-fired power plants. Other power generation technologies such as hydropower were affected too.

The gas turbine began to make a major impact during the 1980s as an engine for power stations. The machine was perfected during and after World War II as an aviation power unit but soon transferred to the power industry for use in power plants supplying peak demand.

During the 1980s the first large base-load power stations using both gas turbines and steam turbines, in a configuration known as the combined cycle plant, were built. This configuration has become the main source of new base-load generating capacity in many countries where natural gas is readily available.

The first years of the twenty-first century have seen renewed emphasis on new and renewable sources of electricity. Fuel cells, a technically advanced but expensive source of electricity, are approaching commercial viability. There is renewed interest in deriving energy from oceans, from waves and currents, and from the heat in tropical seas. Offshore wind farms have started to multiply around the shores of Europe.

The story of the twenty-first century is likely to be the contest between these new technologies and the old combustion technologies for dominance within the power generation industry. And while they battle for supremacy there remains one technology, nuclear fusion, which has yet to prove itself but just might sweep the board.

The politics of electricity

During the last years of the nineteenth century, when the technology was in its infancy, the generation of electricity was seen as one more opportunity for entrepreneurs and joint stock companies to make money. After all, electricity was not unique. There were other means of delivering energy; district heating was already common in the USA and in some European cities while hydraulic power was sold commercially in cities such as London.

As a consequence the early history of the electricity industry was one of small, privately owned companies. Gradually, however, the distribution of electricity rendered most other ways of distributing energy across a network obsolete.

In the twentieth century, as the primacy of electricity became obvious, the distribution of electricity gradually became seen as a public service. Like water, sewage and later gas supply, electricity was needed to operate a modern civilisation. In much of the world, the electricity industry was absorbed by government and became publicly owned. In countries such as the USA where this did not happen, legislation was introduced to govern the supply.

In the late twentieth century, political ideologies changed. Government ownership of industry, including the electricity industry, began to be seen as unnecessary and uneconomic. A move began to convert publicly owned utilities into privately held companies. Alongside this, utility legislation was relaxed to open electricity markets to competition.

By the beginning of the twenty-first century this had become a global phenomenon. A few centralised governments still retained full control over their electricity industries but most paid at least lip service to the concept of liberalisation.

Liberalisation has resulted in both successes and failures. California recorded the most dramatic failure when liberalisation resulted in a virtual breakdown of its electricity supply system, with almost catastrophic consequences. The cost of electricity in California rose dramatically as a result. Elsewhere prices fell after liberalisation.

If state control of the electricity industry was seen to be overbearing and too rigid, a liberalised industry may have too much freedom. Economic rather than political considerations are paramount. This makes government policy more difficult to implement.

Renewable energy offers a good example. A government that wants to increase the proportion of electricity generated from renewable sources cannot simply pass an order down the line. It must use taxes and systems of allowances and penalties; generating companies may chose to pay the penalties if that is the most economically attractive option. In that case the desire of government is ignored.

It is impossible to predict whether modern-free market rules will continue to dominate the electricity industry. Life is full of ironies; instances of policies that are turned on their head by one generation and then turned again a generation later are far from rare. It would be hasty to assume that this will not happen in the utility industries.

The size of the industry

How big is the electricity industry? Tables 1.1 and 1.2 provide the answer. Table 1.1 shows the amount of electricity generated across the globe in 2000. Production is broken down in the table both by region and by type.

Gross electricity generation in 2000 was 14,618 TWh. This is equivalent to roughly 1,670,000-MW power stations running continuously for a year. In fact, the actual global installed capacity in 2000 was over twice that, 3,666,000 MW.[1]

When generation is broken down by type, thermal generation is seen to be dominant. This category refers to power generated from coal, oil or gas. These three fuels were responsible for 9318 TWh, 64% of all the electricity generated in 2000. Hydropower was the next most important source, providing 2628 TWh (18%) with nuclear power a close third (2434 TWh, 17%).

Table 1.1 *World electricity production (in TWh), 2000*

	Thermal power	*Hydro power*	*Nuclear and other power*	*Geothermal power*	*Total*
North America	2997	658	830	99	4584
Central and South America	204	545	11	17	777
Western Europe	1365	558	849	75	2847
Eastern Europe and former USSR	1044	254	266	4	1568
Middle East	425	14	0	0	439
Africa	334	70	13	0	417
Asia and Oceana	2949	529	465	43	3986
Total	9318	2628	2434	238	14,618

Source: US Energy Information Administration.[2]

Table 1.2 *World electricity generating capacity (in GW), 2000*

	Thermal power	Hydro power	Nuclear and other power	Geothermal power	Total
North America	662	176	110	17	965
Central and South America	68	115	3	3	189
Western Europe	360	147	128	14	648
Eastern Europe and former USSR	299	80	49	0	428
Middle East	97	4	0	0	101
Africa	82	20	2	0	104
Asia and Oceana	684	171	70	5	930
Total	2252	713	362	39	3366

Source: US Energy Information Administration.[3]

Regionally, North America produced the largest amount of electricity in 2000, followed by Asia and Oceana. The most striking regional figure is that for African production, 417 TWh or less than one-tenth that of North America. Central and South America also has an extremely low output, 777 TWh. If one wants to identify the poorest regions of the world, one needs to look no further than this table.

Table 1.2 provides figures for the actual installed generating capacity which existed across the globe in 2000. The figures here broadly mirror those in Table 1.1, but there are one or two features to note.

Firstly global nuclear capacity is only half that of global hydropower capacity but contributes almost as much electricity. This reflects the fact that hydropower plants cannot run at 100% capacity throughout the year because they depend on a supply of water and this will vary from season to season. Nuclear power plants, by contrast, work best if they are always operated flat out.

Secondly the gross capacity, 3366 GW is twice as much generating capacity as is required to generate the electricity in Table 1.1, if every station was running flat out all the time. Clearly many plants are working at less than half capacity. We have already seen that hydropower cannot run at full capacity. There will, in addition, be spare capacity in many regions of the world that is only called on during times of peak demand.

We might also note, as both tables indicate, that Central and South America rely on a renewable source, hydropower, for the majority of their electricity. In every other region of the world, thermal power plants are dominant. The composition of the world's power generating capacity is not likely to remain static. New types of generation are becoming ever more competitive and these can be expected to prosper as the present century advances. Renewable technologies, in particular, will advance as environmental concerns and the cost of fossil fuels restrict the use of thermal power stations. What these advancing technologies are and how they work forms much of the subject matter for the remainder of this book.

End notes

1 International Energy Annual 2001, published in 2003 by the US Energy Information Administration.
2 Refer *supra* note 1.
3 Refer *supra* note 1.

2 Environmental considerations

The power generation industry, taken as a whole, is the world's biggest industry. As such it has the largest effect of any industry on the conditions on earth. Some of the effects, particularly the ones associated with the combustion of fossil fuels, are far-reaching both geographically and temporally.

Awareness of the dangers associated with this and other aspects of power generation has been slow to register but since the 1970s a series of events have provided graphic evidence. Acid rain during the 1980s; nuclear disasters such as Chernobyl in 1986; critical reviews of large hydropower projects in the 1980s and 1990s; the recognition of the dangers of global warming during the 1990s; the ever-present haze that has blighted many of the world's cities for 20 years and more: by the end of the twentieth century concern for the environment was one of the major international issues.

As a consequence of this, environmental concerns are beginning to shape the power generation industry. This is an effect that will continue through the next four or five decades of the twenty-first century.

There are environmental considerations which relate to each different type of power generation technology. These are considered in turn, in conjunction with the technologies, in the chapters that follow. This chapter takes a broader look at the most important issues and how they are being treated.

The evolution of environmental awareness

Man has always changed his surroundings. Some of those changes we no longer even recognise; the clearing of forests to create the agricultural farmlands of Europe for example. No one now sees these fields as forests that once were.

Similar changes elsewhere are more obviously detrimental to local or global conditions. Tropical rain forests grow in the poorest of soils. Clear them and the ground is of very little use. Not only that, but the removal of forest cover can lead to erosion, and flooding as well as the loss of groundwater. Most of these effects are negative.

Part of the problem is the ever-increasing size of the human population. Where native tribes could survive in the rain forests in Brazil, the encroachment of outsiders has led to their erosion.

A similar effect is at work in power generation. When the demand for electricity was limited, the effect of the few power stations needed to supply that demand was small. But as demand has risen, so has the cumulative effect. Today that effect is of such a magnitude that it can no longer be ignored.

Consumption of fossil fuel is the prime example. Consumption of coal has grown steadily since the industrial revolution. The first sign of trouble resulting from this practice was the ever-worsening pollution in some major cities. In London the word smog was invented at the beginning of the twentieth century to describe the terrible clouds of fog and smoke that could remain for days. Yet it was only in the 1950s that legislation was finally introduced to control the burning of coal in the UK capital.

Consumption of coal still increased but with the use of smokeless fuel in cities and tall stacks outside, problems associated with its combustion appeared to have been solved. Until, that is, it was discovered that forests in parts of northern Europe and North America were dying and lakes were becoming lifeless. During the 1980s the cause was identified; acid rain resulting from coal combustion. More legislation, aimed at controlling the emission of acidic gases such as sulphur dioxide and nitrogen oxides, was introduced.

Acid rain was dangerous but worse was to come. By the end of the 1980s scientists began to fear that the temperature on the surface of the earth was gradually rising. This has the potential to change conditions everywhere. Was this a natural change or man-made? Scientists did not know.

As studies continued, evidence suggested that the effect was, in part at least, man-made. The rise in temperature followed a rise in the concentration of some gases in the atmosphere. Chief among these was carbon dioxide. One of the main sources of extra carbon dioxide was the combustion of fossil fuels such as coal.

If this is indeed the culprit, and it would appear prudent to assume that it is, then consumption of fossil fuel must fall, or measures must be introduced to remove and secure the carbon dioxide produced. Both are expensive. It has now become one of the main challenges for governments all over the world to reduce the amount of carbon dioxide being released into the atmosphere without crippling their economies.

The way in which fossil fuel is used in power generation is gradually changing as a result of these discoveries and the legislation that has accompanied them. Other technologies also face challenges. Nuclear power is considered by some to be as threatening as fossil fuel combustion, though it still has its advocates too. Hydropower has attracted bad publicity in recent years but should still have an important part to play in future power generation. Meanwhile there are individuals and groups prepared to go to almost any lengths to prevent the construction of wind farms which they consider unsightly.

Electricity is vital to modern living. One can fairly argue that the modern world is a result of the discovery and exploitation of electricity. Therefore unless the world is going to regress technically the supply of electricity must continue and grow. On that basis, compromises must be sought and technical solutions must be found such that growth does not result in irrevocable damage. These are the challenges that the power industry faces, and with it the world.

The environmental effects of power generation

Much human activity has an effect on the environment and, as already outlined above, power generation is no exception. Some of these effects are more serious than others. The atmospheric pollution resulting from coal, oil and gas combustion has had obvious effects. But combustion of fossil fuel also releases a significant amount of heat into the environment, mostly as a result of the inefficiency of the energy conversion process. Is this a serious side effect? In most cases, it probably is not.

Power stations have a physical presence in the environment. Some people will consider this a visual intrusion. Most make noises, another source of irritation. There are electromagnetic fields associated with the passage of alternating currents through power cables. A power plant needs maintaining, servicing and often needs providing with fuel. That will generate traffic.

Clearly some of these effects are more far-reaching than others. Even so, the local effects of a power station may be a significant issue for the immediately adjacent population. Deciding what weight must be given to such considerations when planning future generating capacity can be a fearsomely difficult issue. It is the big issues, however, particularly global warming, which will have the most significant effect on the future of power generation.

The carbon cycle and atmospheric warming

The combustion of fossil fuels such as coal, oil and natural gas releases significant quantities of carbon dioxide into the atmosphere. Since the industrial revolution the use of these fuels has accelerated. The consequence appears to have been a gradual but accelerating increase in the concentration of carbon dioxide within the earth's atmosphere.

Before the industrial revolution the concentration of carbon dioxide in the earth's atmosphere was around 270–280 ppm. Between 1700 and 1900 there was a gradual increase in atmospheric concentrations but from 1900 onwards the concentration changed more rapidly as shown in Table 2.1. From 1900 to 1940 atmospheric carbon dioxide increased by around

Table 2.1 *Atmospheric carbon dioxide concentrations*

Year	*Carbon dioxide concentration (ppm)*
1700	270–280
1900	293
1940	307
1980	339
2000	369
2050	440–500
2100	500–700

10 ppm, from 1940 to 1980 it increased by 32 ppm and by 2000 it had increased by a further 30 ppm. By then the total concentration was 369 ppm, an increase of over 30% since 1700.

If the increase in carbon dioxide concentration is a direct result of the combustion of fossil fuel then it will continue to rise until that combustion is curbed. Estimates of future concentrations are at best speculative but Table 2.1 includes a range of estimates for both 2050 and 2100. The worst case in Table 2.1 shows concentrations doubling in 100 years.

There is a further caveat. While the evidence for a fossil fuel connection with the increase in concentration of carbon dioxide is compelling, the cycling of carbon between the atmosphere, the sea and the biosphere is so complex that it is impossible to be certain how significant the man-made changes are.

The atmospheric emissions of carbon from human activities such as the combustion of coal, oil and natural gas amount to a total of around 5.5 Gtonnes each year. While this is an enormous figure, it is tiny compared to the total carbon content in the atmosphere of 750 Gtonnes.

This atmospheric carbon is part of the global carbon cycle. There are roughly 2200 Gtonnes of carbon contained in vegetation, soil and other organic material on the earth's surface, 1000 Gtonnes in the ocean surfaces and 38,000 Gtonnes in the deep oceans.

The carbon in the atmosphere, primarily in the form of carbon dioxide, is not static. Plants absorb atmospheric carbon dioxide during photosynthesis, using the carbon as a building block for new molecules. Plant and animal respiration on the other hand, part of a natural process of converting fuel into energy, releases carbon dioxide to the atmosphere. As a result there are probably around 60 Gtonnes of carbon cycled between vegetation and the atmosphere each year while a further 100 Gtonnes is cycled between the oceans and the atmosphere by a process of release and reabsorption. Thus the cycling of carbon between the atmosphere and the earth's surface is a complex exchange into which the human contribution from fossil fuel combustion is small.

The actual significance of the additional release of carbon dioxide resulting from human activity depends on the interpretation of various scientific observations. The most serious of these relate to a slow increase in temperature at the earth's surface. This has been attributed to the greenhouse effect, whereby carbon dioxide and other gases in the atmosphere allow the sun's radiation to penetrate the atmosphere but prevent heat leaving, in effect acting as a global insulator.

If human activity is responsible for global warming, then unless carbon dioxide emissions are controlled and eventually reduced, the temperature rise will continue and probably accelerate. This will lead to a number of major changes to conditions around the globe. The polar ice caps will melt, leading to rises in sea level which will inundate many low lying areas of land. Climate conditions will change. Plants will grow more quickly in a carbon-dioxide-rich atmosphere.

Not all scientists agree that changes in our practices can control the global changes. There have been large changes in atmospheric carbon dioxide concentrations in the past, and large temperature swings. It remains plausible that both carbon dioxide concentration changes and global temperature changes are part of a natural cycle and that the human contribution has little influence.

It may be impossible to find absolutely conclusive proof to support one argument or the other. But in the meantime conditions will continue to change. And if human activity is responsible, the change may eventually become irreversible. Besides, it is clear that combustion of fossil fuel is creating more carbon dioxide than would naturally have been available. The sound environmental response is to stop this man-made change to global conditions.

Controlling carbon dioxide

Fossil fuels are all derived from biomass, from trees and vegetation which grew millions of years ago and subsequently became buried beneath the surface of the earth. Without man's intervention the carbon contained in these materials would have remained buried and removed from the carbon cycle. As a result of human activity they have been returned to the carbon cycle.

An immediate cessation of all combustion of fossil fuel would stabilise the situation. That is currently impossible. The popular strategy of switching fuel from coal to gas reduces the amount of carbon dioxide generated but does not eliminate it.

One short-term measure would be to capture the carbon dioxide produced by a combustion power station and store, or sequester it in a way that would prevent it ever entering the atmosphere. Technologies exist that are capable of capturing the carbon dioxide from the flue gas of a power plant. Finding somewhere to store it poses a more difficult problem.

One solution is to pump it into exhausted oil and gas fields. There are other underground strata in which it might be stored. A third possibility is to store it at the bottom of the world's oceans. The enormous pressures found there would solidify the gas and the solid would remain isolated unless disturbed.

These solutions are all expensive and none is particularly attractive. However they may become necessary as short-term solutions. Over the longer term the replacement of fossil fuels with either renewable technologies that do not rely on combustion or with biomass generated fuel which releases carbon dioxide when burnt but absorbs it again when it is regrown, will be necessary.

The hydrogen economy

A switch to sustainable renewable technologies would appear to be the most practical means to control power plant emissions of carbon dioxide but it will not solve all global problems associated with fossil fuel. What about all the other uses, particularly for automotive power?

A more radical solution would be to switch from an economy based on fossil fuel to one based on hydrogen. Fossil fuels, particularly oil and gas, have become a lynch pin of the global economy because they are so versatile. The fuels are easily stored and transported from one location to another. They can be used in many different ways too; power stations, internal combustion engines, cookers, refrigerators, all these and more can be powered will one of these fuels.

Renewable electricity sources such as hydropower, solar power, wind power and biomass can replace fossil fuel in power generation but they cannot easily be adapted to meet all the other uses to which fossil fuel is put. When they can, the solution is often more expensive and less convenient than the fossil fuel alternative.

Hydrogen, on the other hand can replace fossil fuel in virtually all applications. It can be used to power an internal combustion engine. It can be burnt to provide heating or cooling. Moreover it can be stored and transported with relative ease. And it is clean. When it is burnt, the only product of its combustion is water.

Where would the hydrogen for a hydrogen economy be found? The primary source would be water and the best way of making it would be by use of electrolysis. Renewable energy power plants would generate electricity and the electricity would be used to turn water into hydrogen and oxygen. The hydrogen would be captured and stored for future use.

This may seem an expensive and inefficient method of generating fuel. Indeed it is. For the hydrogen economy to work, electricity from renewable sources needs to become much cheaper, cheaper probably than all but the cheapest electricity today. Even so it offers a vision for the future in which life continues in much the same way as it does today.

Externalities

What, exactly, is the cost of electricity today? That is not an easy question to answer. In basic economic terms the cost depends on the cost of the power station – how much is cost to build[1] – the cost of operating and maintaining it over its lifetime which is typically 30 years for a combustion plant, and the cost of fuel. If all these numbers are added up and divided by the number of units of electricity the plant produces over its lifetime, then this is the basic cost of electricity. (In most cases there will be an addition to this basic cost to cover either profits if the plant is owned by a private company or for future investment if the plant is publicly owned.)

Practically this means of accounting is impossible because the plant has to start selling power as soon as it is capable of operating. So some guesses will have to be made about its future performance and costs over its lifetime and some economic factors will be needed to take account of the changing value of money. While these add some uncertainty, the basic equation remains the same.

At the beginning of the twenty-first century, using this equation, the new power plant offering the cheapest source of electricity appears to be the gas-fired combined cycle power station. It is cheap and quick to build and relatively easy to maintain. The fuel is the most significant determinant of electricity price, so while gas is cheap, so is electricity.

But does the basic economic equation take account of all the factors involved in generating electricity? There is a small but growing body of opinion which says no. It says that there are other very important factors that need to be taken into account. These are generally factors such as the effect of power production on the environment and on human health, factors which society pays for but not the electricity producer or consumer directly. These factors are called *externalities*.

A major study carried out by the European Union (EU) and the USA over a decade in the 1990s estimated that the cost of these externalities, excluding the cost of global warming, were equivalent to 1–2% of the EU's Gross Domestic Product.

The cost of electricity in the EU in 2001, when the report of the study was published, was around €0.04/kWh. External costs for a variety of traditional and renewable energy technologies as determined by the study are shown in Table 2.2. Actual external costs vary from country to country and the table shows the best and worst figures across all countries. These figures indicate that coal combustion costs at least and additional €0.02/kWh on top of the €0.04/kWh paid by the consumer. Gas-fired

Table 2.2 *External cost of power generation technologies*

	Cost (€/kWh)
Coal and lignite	2–15
Peat	2–3
Oil	3–11
Gas	1–4
Nuclear	0.2–0.7
Biomass	0.2–3
Hydropower	0–1
Photovoltaic	0.6
Wind	0.1–0.3

Source: Figures are from the ExternE project funded by the European Union and the USA.

generation costs an additional €0.01/kWh while the external costs for nuclear and most renewable technologies are a fraction of this.

If consumers were forced to pay these external costs – by the imposition by governments of some form of surcharge for example – the balance in the basic equation to determine the cost of electricity would shift in favour of all the non-combustion technologies. Of course such a move would initially penalise consumers because a high proportion of the world's electricity comes from fossil fuel and the capacity cannot be replaced overnight. It would not suit fossil fuel producers either and it would drive up manufacturing prices, initially at least, affecting global economics. Over the long term, if the analysis is correct, everyone would benefit.

Life-cycle assessment

Another important tool for establishing the relative performance of power generation technologies is a life-cycle assessment. The aim of a life-cycle assessment is to measure the performance of a power plant with reference to one or more parameters such as its emissions of carbon dioxide or the energy efficiency of its power generation. The assessment covers the complete life of the plant starting from the manufacture of the components that were used to construct it and ending with its decommissioning.

Thus if one were examining the amount of carbon dioxide produced by a gas-fired power plant one would examine not only the amount produced by burning gas in the plant but also that produced when electricity or some other form of energy was used to manufacture the components used to build the plant, and any used when the plant was dismantled and recycled.

Table 2.3 *Lifetime missions of carbon dioxide for various power generation technologies*

	Carbon dioxide emissions (tonnes/GWh)
Coal	964
Oil	726
Gas	484
Nuclear	8
Wind	7
Photovoltaic	5
Large hydro	4
Solar thermal	3
Sustainable wood	−160

Source: European Union.[2]

If carbon dioxide emission is studied, figures show that coal-, gas- and oil-fired power plants produce massively more carbon dioxide for each unit of electricity they produce that do most renewable technologies. Typical figures are given in Table 2.3. Similar results are found for other common combustion plant emissions such as sulphur dioxide, carbon monoxide and nitrogen oxides.

One of the most interesting types of life-cycle assessment is the total energy balance of a plant, a figure which indicates how many units of energy a power station produces for each unit of energy it consumes over its lifetime. Energy is used to manufacture components for a power plant. Energy is needed to produce and deliver combustion fuel to a power plant. The fuel itself contains energy which is consumed. And energy is consumed during the decommissioning of a power plant. All these units of energy must be added together and then divided by the total number of units of energy the power station delivers during its lifetime to provide the energy balance.

It is probably self-evident that most renewable technologies will score more highly than fossil fuel power plants on the measure, though the manufacture of solar cells is relatively energy intensive. What is not so obvious is that burning biomass is actually significantly more energy efficient than the combustion of coal or gas even though all are combustion processes. Yet analysis suggests that the energy balance of a biomass plant is three times better than a coal-fired plant and almost six times better than a gas-fired plant (for more detail see Chapter 15). Much of the difference results from the energy used to mine and transport coal and gas; gas loses during transport are also a significant factor.

The bottom line

Most environmental assessments of power generation indicate that there are environmental benefits to be gained in shifting from reliance on fossil fuels to other, primarily renewable, forms of generation. In most cases, however, the determining factor remains cost. Indeed cost has become more decisive over the last 20 years as the control over the power generation industry has shifted, in many parts of the world, from the public sector to the private sector. Economics do not always favour the most environmentally favourable solutions.

The private sector requires short-term return on investment. This favours technologies that are cheap to build because loans for construction are small and can be repaid quickly. Most renewable technologies are capital intensive. The generating plant costs a lot to build but very little to run because the fuel – be it wind, sunlight or water – is usually free. These plants are more cost effective over the long term, probably 20 years or longer, but less so over a shorter term.

Governments cannot direct the private sector but it can influence the industry with legislation, surcharges and incentives. Such governmental tools are being used with some effect. Financial institutions are also beginning to heed the shift in consensus. In June 2003 a group of commercial banks agreed a set of guidelines called the *Equator Principles* which are intended provide a framework for assessing the social and environmental issues associated with a project seeking a loan. These guidelines are voluntary but potentially significant.

A shift away from fossil fuels will have a profound effect on the whole power generation industry. Not only generation but transmission and distribution management and structure will be affected. The change will, initially at least, be expensive. As a result change will come slowly. What does appear clear, at the beginning of the twenty-first century, is that the change will come.

End notes

1 The cost of construction of the power plant must also include the cost of servicing any loans raised in order to build the plant. These may end up the most important factor of all, particularly where commercial loans have to be raised for a power plant being constructed by a private company.
2 Concerted Action for Offshore Wind Energy in Europe, 2001. This is a European Commission supported report published by the Delft University Wind Energy Research Institute.

3 Coal-fired power plants

Coal is the world's most important and the most widely used fuel for generating electricity. According to the World Energy Council it provides 23% of total global primary energy demand and 38% of electricity production.[1] Total world production of coal in 1999 was 4,343,151,000 tonnes, and consumption was 4,409,815,000 tonnes.

The importance of coal is reinforced by national statistics from the main global consumers. In the USA, coal-fired plants produce 51% of the nation's power. This dominance is expected to continue well into the twenty-first century. In China, coal-fired stations were generating 65% of the electricity in 1988, and by beginning of the twenty-first century 75% of the country's electricity came from fossil fuel, mostly coal. In India too, fossil fuel, again primarily coal, accounts for around 71% of installed capacity.

The major attraction of coal is its abundance. Significant deposits can be found in most parts of the world, from the USA to South Africa, across Europe, in many parts of Asia and in Australia. Exceptions exist, such as Japan and Taiwan, where resources are limited; these countries import vast quantities of coal. Among the continents, only South America and Africa – outside South Africa – have limited reserves.

According to the World Energy Council's 2001 Survey of Energy Resources, the proved recoverable world resources of bituminous coals, sub-bituminous coals and lignites amount to 984,453 Mtonnes. (Anthracite, the hardest coal, is rarely used for power generation when alternatives are available.) Figures for these reserves, broken down by coal type and by region, are given in Table 3.1.

Figures for proved reserves, such as those in Table 3.1, reflect the extent to which a resource has been surveyed rather than offering a measure of the actual amounts of coal that exist. Potential reserves greatly exceed the identified reserves, and estimates of the latter are usually conservative. At current consumption levels, proved reserves of coal can continue to provide energy for at least 200 years.

Coal is the cheapest of fossil fuels, another reason why it is attractive to power generators. However it is expensive to transport, so the best site for a coal-fired power plant is close to the mine that is supplying its fuel.

Coal is also the dirtiest of the fossil fuels, producing large quantities of ash, sulphurous emissions, nitrogen oxides (NO_x) emissions and carbon dioxide, and releasing significant concentrations of trace metals. As a result the combustion of coal has been responsible for some of the worst

Table 3.1 *Proved global coal reserves*

	Bituminous (Mtonnes)	*Sub-bituminous (Mtonnes)*	*Lignite (Mtonnes)*	*Total*
Africa	55,171	193	3	55,367
North America	120,222	102,375	35,369	257,966
South America	7738	13,890	124	21,752
Asia	179,040	38,688	34,580	252,308
Europe	112,596	119,109	80,981	312,686
Middle East	1710	–	–	1710
Oceania	42,585	2046	38,033	82,664
Total	519,062	276,301	189,090	984,453

Source: World Energy Council, Survey of Energy Resources 2001.

environmental damage, barring accidents, created by heavy industry anywhere in the world.

In consequence, coal has developed a bad environmental image. But developments since the 1980s aimed at controlling emissions from coal-fired plants, combined with new coal-burning technologies, mean that a modern coal-fired power plant can be built to meet the most stringent environmental regulations, anywhere in the world. Techniques for capturing sulphur, nitrogen emissions and ash are well established. The next challenge it to develop cost-effective ways of removing and storing carbon dioxide, for of all fossil fuels, coal produces the largest quantity of this greenhouse gas.

Modern coal-fired power plants, with emission-control systems, are more expensive than the older style of plant common before the mid-1980s. Even so, coal remains the cheapest way of generating power in many parts of the globe. Whatever the environmental constraints, the fuel will continue to provide a substantial proportion of the world's electricity for much of the coming century.

Types of coal

The term coal embraces a range of materials. Within this range there are a number of distinct types of coal, each with different physical properties. These properties affect the suitability of the coal for power generation.

The hardest of coals is anthracite. This coal contains the highest percentage of carbon (up to 98%) and very little volatile matter. As a result, anthracite from many sources is slow burning and difficult to fire in a power station boiler unless it is mixed with another fuel. There are large reserves of anthracite around the world, particularly in Asia. In consequence, power plants are being built to burn this fuel alone.

Though anthracite is abundant, the largest group of coals are the bituminous coals. These coals contain significant amounts of volatile matter. When they are heated they form a sticky mass, from which their name is derived. Bituminous coals normally contain above 70% carbon. They burn easily, especially when ground or pulverised. This makes them ideal fuels for power stations. Bituminous coals are further characterised, depending on the amount of volatile matter they contain, as high-, medium- or low-volatile bituminous coals.

A third category, sub-bituminous coals, are black or black-brown. These coals contain between 15% and 30% water, even though they appear dry. They burn well, making them suitable as power plant fuels.

The last group of coals that are widely used in power stations are lignites. These are brown rather than black and have a moisture content of 30–45%. Lignites are formed from plants which were rich in resins and contain a significant amount of volatile material. The amount of water in lignite, and its consequent low carbon content, makes the fuel uneconomic to transport over any great distance. Lignite-fired power stations are usually found adjacent to the source of fuel.

A type of unconsolidated lignite, usually found close to the surface of the earth where it can be strip-mined, is sometimes called *brown coal*. (This name is common in Germany.) Brown coal has a moisture content of around 45%. Peat is also burned in power plants, though rarely.

Coal cleaning and processing

Coal cleaning offers a way of improving the quality of a coal, both economically and environmentally. The most well-established methods of coal-cleaning focus on removing excess moisture from the coal and reducing the amount of uncombustible material which will remain as ash after combustion. Moisture removal reduces the weight and volume of the coal, rendering it more economical to transport. Ash removal improves its combustion properties and aids power plant performance.

Moisture can be removed from coal by drying. This can simply be solar drying, leaving the coal in the open before transporting it. The alternative, producing heat to dry the coal, is a more expensive option.

Drying coal by heating is most often carried out at the power station, utilising surplus energy in the plant flue gases. Such a procedure is absolutely essential when burning high-moisture lignites such as brown coal. It does not, however, affect the transportation costs because the fuel has, by this stage, already reached the power station.

Ash removal is carried out by crushing the coal into small particles. Incombustible mineral particles are more dense than the coal and can be separated using a gravity-based method. Such treatment will remove some minerals containing sulphur, and can result in a reduction of up to 40% in sulphur dioxide emissions during combustion. (Some sulphur is bound to the carbon in the coal. Such sulphur is not affected by this type of cleaning.)

There have been attempts to develop more advanced methods for coal treatment employing either higher-temperature processing of the coal or chemical rather than physical processes. These have not, so far, found commercial application.

According to the World Bank,[2] the cost of cleaning coal using existing technology is between US$1/tonne and US$10/tonne depending on the degree of cleaning required. This type of technology is simple and can be deployed in most parts of the world.

Traditional coal-burning power plant technology

The modern technology used for burning coal to generate electricity has evolved over a period of more than a century and until awareness grew of the environmental damage coal burning could produce, the coal-fired power station developed in a single direction.

The basic principle underlying this type of power station is to burn coal in air and capture the heat released to raise steam for driving a steam turbine. The rotation of this steam turbine, in turn, drives a generator; the net result is electricity.

The traditional coal-fired power plant comprises two basic components. The first component is a furnace boiler designed to burn the coal and capture the heat energy released using a system of circulating water and steam. The second part of the system is a steam turbine generator which

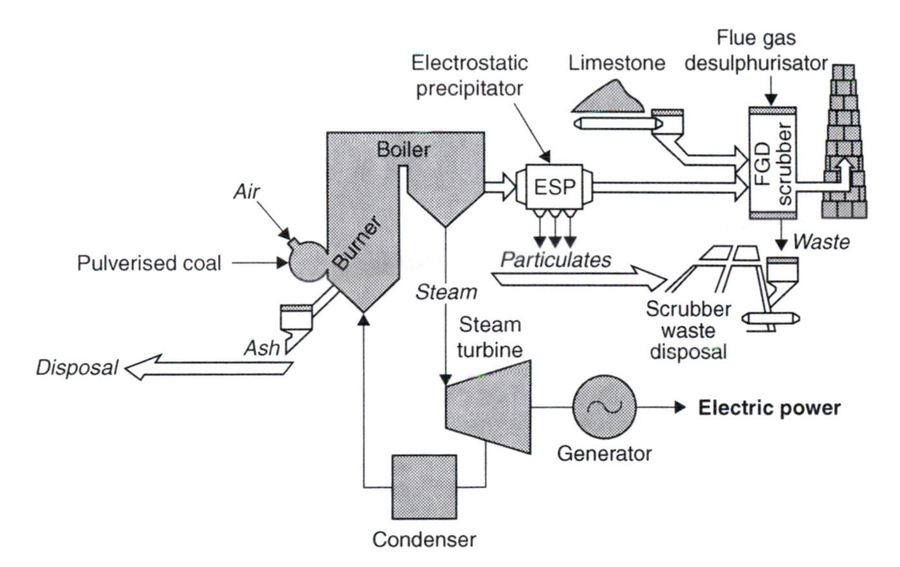

Figure 3.1 *Flow diagram of a traditional coal-fired power plant*

converts the heat energy captured by the steam into electrical energy. In other words, chemical energy held within the coal is first translated into heat energy and then into mechanical energy, and finally into electrical energy. Modern plants also include additional units to remove dust and acid emissions from the flue gases before they are released into the atmosphere.

Boiler technology

A power plant boiler is a device for converting the chemical energy in coal into heat energy and then transferring that heat energy to a fluid, steam. The efficiency of a coal-fired power plant increases as the pressure and temperature of the steam increases. This has led to a demand for higher temperatures and pressures as technology has developed and this has required, in turn, the development of materials with higher performance under increasingly stressful conditions. The most advanced boilers develop steam with a pressure of around 250 bar and a temperature of 600°C.

Early boilers were made from iron, but as the demands on the system increased, special steels were used that could resist the conditions encountered in the power plant. These now dominate in modern boilers. Even so, oxygen dissolved in the water circulating within the boiler pipes can cause serious corrosion in steel at the elevated temperatures and pressures to which it is exposed, so the boiler water must be deoxygenated.

The first part of the boiler is a furnace in which combustion takes place. In the most common type of boiler, pulverised coal is injected with a stream of air into the furnace in a continuous process through a device known as a *burner*. The coal burns, producing primarily carbon dioxide while incombustible mineral material (ash) falls to the bottom of the furnace where it can be removed (some is also carried away by the hot combustion gases).

The heat generated during combustion (the temperature at the heart of the furnace may be as high as 1500°C) is partly radiant and partly convective, the latter carried off by the hot combustion gases. The radiant heat is collected at the walls of the furnace where water is circulated in pipes.[3] Covective heat in the combustion gases is captured in bundles of tubes containing either water or steam which are placed in the path of the flue gas as it exits the furnace.

In a conventional boiler there is a drum positioned appropriately within the steam–water system containing both water and steam so that steam can develop as the temperature of the fluid rises. The most advanced designs, however, operate at such high temperatures and pressures that they do not pass through a stage in which water and steam co-exist. In these boilers the water turns directly to steam within the watertubes. This type of boiler exploits what is called a *supercritical cycle*, called so because the thermodynamic fluid (the water) enters what is known as the *supercritical phase* without passing through a condition in which both water and steam co-exist.

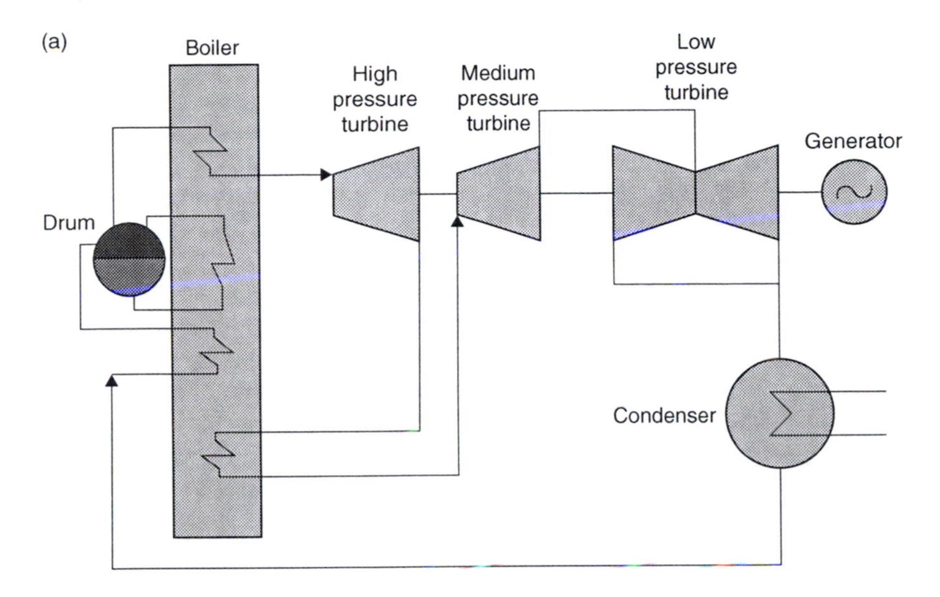

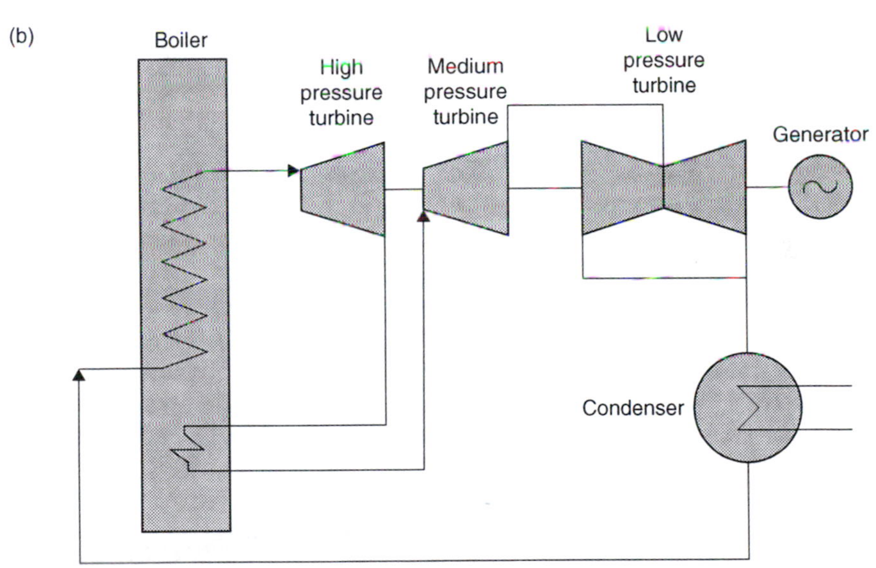

Figure 3.2 *Coal-fired power station boiler steam cycles: (a) typical subcritical steam cycle with a conventional drum boiler and natural circulation and (b) typical supercritical steam cycle with once-through boiler*

The boiler watertubes in the exhaust gas path are frequently divided into a number of different sections. (These sections have names, such as economiser, reheater or superheater.) The water or steam passes through them is a specific order determined by the design of the steam cycle. Traditional pulverised-coal boilers have been built with outputs of up to 1000 MW.

Steam turbine design

The steam turbine is the primary mechanical device in most conventional coal-fired power stations. Its job is to convert the heat energy contained in the steam exiting the boiler into mechanical energy, rotary motion. The steam turbine first appeared in power applications at the end of the nineteenth century. Before that steam power was derived from steam-driven piston engines.

The steam turbine is something of a cross between a hydropower turbine and a windmill. It, like them, is designed to extract energy from a moving fluid. The fluid is water, the same as the hydro turbine. In the case of a hydro turbine the water remains in the liquid phase and neither its volume nor its temperature changes during energy extraction. In the case of the steam turbine, energy extraction is from a gas, steam, rather than a liquid and involves both the pressure and the temperature of the fluid falling. This has a profound effect on the turbine design.

Both hydro and steam turbines exist in two broad types: there are *impulse turbines* which extract the energy from a fast-moving jet of fluid and *reaction turbines* which are designed to exploit the pressure of a fluid rather than its motion. A hydro turbine will be of one design or the other. In a steam turbine the two principles may be mixed in a single machine and they may even be mixed in a single turbine blade.

It is impossible to extract all the energy from steam using a turbine with a single set of turbine blades. Instead, a steam turbine utilises a series of sets of blades, called *stages*. Each stage is followed by a set of stationary blades (usually called *nozzles*) which control the steam flow to the next stage.

A single steam turbine stage consists of a set of narrow blades projecting from a central hub. (In concept, it is something like a steam windmill.) Ten or more sets of blades can be mounted on a single steam turbine shaft. This combination of shaft and blades is called a *rotor*. The turbine stages are separated by carefully designed stationary blades, or nozzles, which control the flow of steam from one set of rotating blades to the next. The precise shape of the blades in each set determines whether that set is impulse or reaction, or a cross between the two. The hub, blades and nozzles are enclosed in a close-fitting case to maintain the steam pressure.

In a steam turbine impulse stage, energy is extracted at constant pressure while the velocity of the steam falls as it flows across the blades. The steam is then expanded through a stationary control stage to increase its velocity again before energy is extracted from another set of impulse blades. In a steam turbine reaction stage, by contrast, both pressure and velocity of the steam fall as energy is extracted by the rotating blades.

Steam exiting the power plant boiler is at a high temperature and a high pressure. Both temperature and pressure fall as the steam passes through the turbine. The greater the temperature drop and the greater the pressure

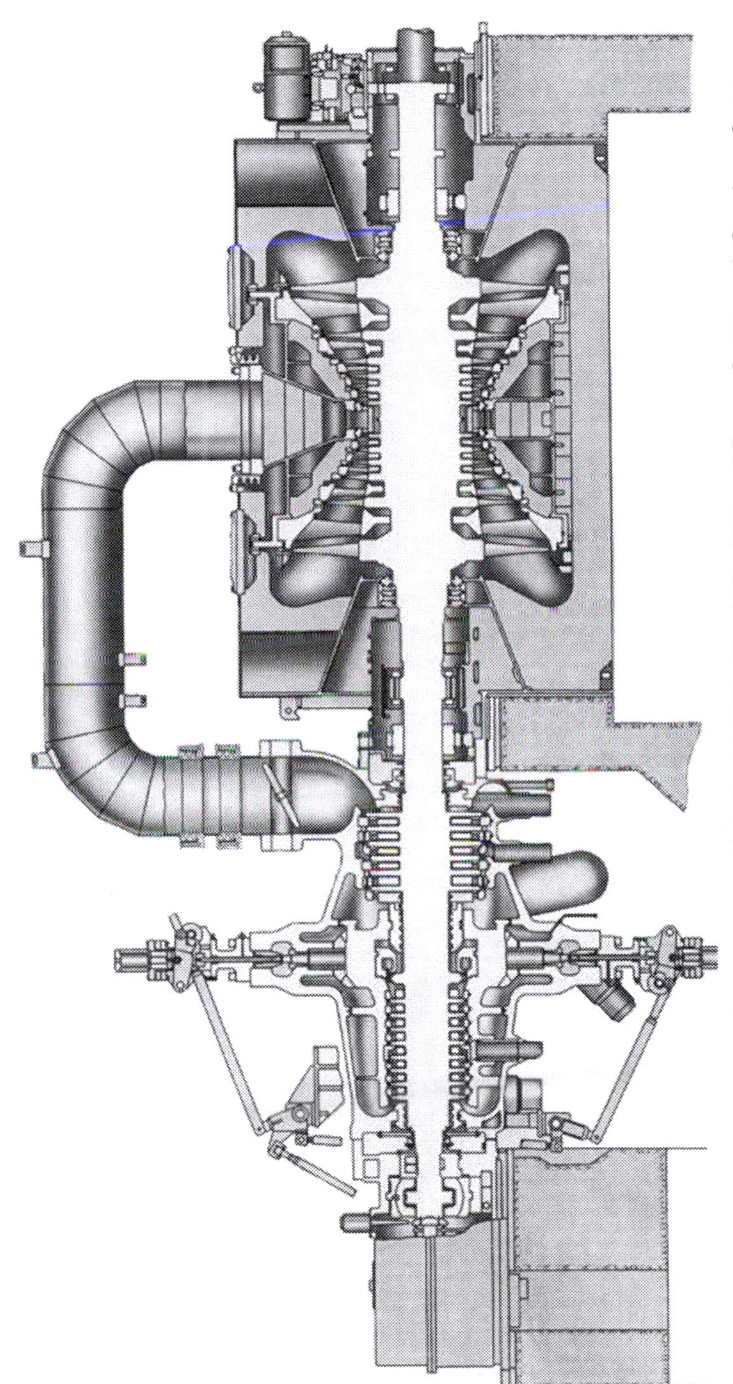

Figure 3.3 *Section through a modern steam turbine.* Source: *Toshiba Industrial and Power Systems & Services Company*

drop, the more energy can be captured from the steam. Consequently the most efficient power plants condense the steam back to water at the end of the turbine.

Even with a modern design it is impossible to capture all the energy from the steam efficiently with a single multiple-stage turbine. Coal-fired power plants use several. These are usually divided into high-, medium- and low-pressure turbines. The blades in these turbines get larger (longer) as the pressure drops; in fact, the low-pressure turbine may comprise several turbines operating in parallel to gain the most energy without making a single turbine impossibly large. All the turbines may be mounted on a single shaft, but it is common for the low-pressure turbines to be on a separate shaft rotating at a lower speed to reduce the forces exerted at the blade tips. Multiple turbines of this type can have aggregate outputs of over 1000 MW.

As with boilers, the demands of modern power plant design have led to the development and introduction of high-performance materials that can cope with the extreme conditions encountered within a steam turbine. The high-pressure turbine blades have to be able to withstand extremes of both temperature and pressure and have to be able to resist the abrasive force of steam. At the low-pressure end of the turbine train the large size of the turbines means that the blade tip speeds are enormous, again requiring specially designed materials to withstand the centrifugal forces exerted on them.

A refinement which improves the overall efficiency is to return the steam to the boiler after it has passed through the high-pressure turbine, reheating it before delivering it to the medium-pressure turbine. Most modern steam turbine plants use this single reheat design (multiple reheat is also possible).

The theoretical maximum efficiency of a coal-fired power station is determined by the temperature difference between the steam entering the high-pressure turbine and the steam exiting the low-pressure turbine. The greater this temperature difference, the more energy can be extracted. With the most advanced technology, utilising the best boiler materials to achieve the highest-steam temperatures and pressures, a maximum efficiency of around 43–45% can be achieved. New supercritical designs may eventually push this as high as 55%. In the near future, however, the best that is likely to be achieved is something between 47% and 49%.

Generators

The turbine shaft, or shafts if there is more than one, are coupled to a generator which converts the rotary mechanical motion into the electrical energy that the plant is designed to provide. Generators, like steam turbines, first appeared during the nineteenth century. All utilise a coil of a conducting material, usually copper moving in a magnetic field to generate electricity.

The generators used in most power stations, including coal-fired power stations, are designed to deliver an alternating current (AC) to a power grid. An AC current is preferred because it allows the voltage to be raised or lowered easily using a transformer. For transmission of power over long distances it is preferable to use a very high voltage and a low current. The voltage is then reduced with a transformer before delivery to the consumer.

The need to generate an AC voltage determines the speed at which the generator rotates. This must be an exact multiple of the grid frequency (normally grids operate at either 50 or 60 Hz). For grids operating at 50 Hz the traditional generator speed is 50 cycles per second, or 3000 rpm. The equivalent 60 Hz machine rotates at 3600 rpm. This speed, in turn, determines the operating speed of the steam turbine. Large low-pressure steam turbines may operate at half these speeds.

Generators may be as large as 2000 MW, and large generators are normally built to suit a particular project. Modern generators operate with an efficiency of greater than 95%. The remaining 5% of the mechanical input energy from the turbine is usually lost as heat within the generator windings and magnetic components. Even though the percentage is small, this still represents an enormous amount of energy; perhaps 50 MW in a 1000-MW machine. Hence generators require very efficient cooling systems in order to prevent them overheating. A variety of cooling mediums are used, including hydrogen which is extremely efficient because of its low density and high specific heat.

The broad outline of generator design has changed little over a century. However new materials have improved efficiencies. The latest developments involve the use of superconducting materials to reduce energy and increase efficiencies.

Emission control for traditional coal-burning plants

The combustion of coal to generate energy is an inherently dirty process. The combustion product is primarily carbon dioxide, one of the main greenhouse gases. High-temperature combustion also produces NO_x, both from nitrogen contained within the coal and from atmospheric nitrogen. If the coal contains sulphur (and all natural coals contain some sulphur), this emerges as sulphur dioxide, a potent chemical that is converted into acid in the atmosphere. Incombustible mineral material in the coal is left as ash and slag which must be disposed of harmlessly. And some mineral and particular material escapes with the flue gases into the atmosphere; this can contain trace metals such as mercury which are potentially harmful.

With such a catalogue of unwanted by-products, it is not surprising that coal combustion has attracted criticism. And as areas that have been laid waste by uncontrolled burning of coal – such as parts of northern India, or areas in eastern Germany testify – such criticism is fully justified.

Modern developments have sought to make coal combustion as environmentally benign as possible. To this end, strategies have evolved to control all the pollutants generated in a coal-fired power plant. These strategies can be extremely effective and while some are costly, others are cheap to implement.

Coal treatment

Cleaning coal prior to combustion can significantly reduce the levels of sulphur emissions from a power plant as well as reducing the amount of ash and slag produced. This can have a beneficial effect on plant performance and maintenance schedules. It has been estimated that boiler availability improves by 1% for every 1% reduction in ash content. The main approach to physical coal cleaning has been outlined above.

One disadvantage of coal cleaning is that it leads to a loss or between 2% and 15% of the coal with the coal waste. However it is possible to burn this coal waste in a fluidised-bed combustion (FBC) plant (see p. 33).

Low nitrogen oxides burners

NO_x are generated by a reaction between oxygen and nitrogen contained in air during combustion. This NO_x production is strongly affected by two factors, the temperature at which the combustion takes place and the amount of oxygen available during combustion. Controlling these parameters provides a way of controlling the quantity of NO_x generated. This is achieved most simply using a low NO_x burner.

A low NO_x burner is a burner which has been designed to create an initial combustion region for the pulverised-coal particles where the proportion of oxygen is kept low. When this happens, most of the available oxygen is captured as carbon dioxide during the coal combustion process, leaving little to react with nitrogen.

To achieve this end, some of the air needed to burn the coal completely is prevented from entering the initial combustion region with the coal; instead it is delayed briefly, being admitted to this primary combustion region after some of the combustion has been completed. This staged combustion procedure (as it is commonly known) can reduce the level of NO_x produced by 30–55%.

The initial combustion zone is normally the hottest region in the furnace. As the combustion gases leave this zone they start to cool. At this stage, further air can be admitted (if combustion of the pulverised coal is still incomplete) to allow the combustion of the fuel to be completed, but at a lower temperature where the production of NO_x is reduced. The air admitted at this stage in the furnace is called *'over-fire-air'*. When used in conjunction with a low NO_x burner, the use of over-fire-air can lead to a reduction in NO_x levels of 40–60%.

A third strategy which can reduce the NO_x level even further is called *reburning*. This simply means that more coal, or natural gas, is introduced into the combustion gases after they have left the combustion zone. The effect is to remove some of the oxides of nitrogen that have been formed. Overall reductions of up to 70% can be achieved. Low NO_x burners, over-fire-air and reburning are all strategies that can be applied to existing coal-fired power plants as well as being incorporated into new plants.

Sulphur dioxide removal

There is no strategy similar to low NO_x burners that can be used to control the emission of sulphur dioxide. If sulphur is present in coal it will be converted into sulphur dioxide during combustion. The only recourse it to capture the sulphur, either before the coal is burnt using a coal-cleaning process, or after combustion using some chemical reagent inside the power plant.

There are many chemicals that are potentially capable of capturing sulphur dioxide from the flue gases of a power station but the cheapest to use are lime and limestone. Both are calcium compounds which will react with sulphur dioxide to produce calcium sulphate. If the latter can be made in a pure enough form it can be sold into the building industry for use in wallboards.

The cheapest method of capturing sulphur dioxide is to inject one of these sorbent materials into the flue gas stream as it exists the furnace. The reaction then takes place in the hot gas stream and the resultant particles of calcium sulphate, and of excess sorbent, are captured in a filter downstream of the injection point.

Depending on the point of injection of the sorbent, this method of sulphur removal can capture between 30% and 90% of the sulphur in the flue gas stream. The cheapest, and least effective method (30–60% capture efficiency) is to inject the sorbent directly above the furnace. Injection later in the flue gas stream is more expensive but can remove up to 90% of the sulphur.

Sorbent injection into a flue gas stream is the cheapest way of capturing sulphur but it is not the most efficient. The best-established method of removing most of the sulphur from the flue gas of a power plant is with a flue gas desulphurisation (FGD) unit, also called a *wet scrubber*.

The FGD unit comprises a specially constructed chamber through which the flue gas passes. A slurry of water containing 10% lime or limestone is sprayed into the flue gas where it reacts, capturing the sulphur dioxide. The slurry containing both gypsum and unreacted lime or limestone is then collected at the bottom of the chamber and recycled.

Typical wet scrubbing systems can capture up to 97% of the sulphur within the flue gas. With special additives, this can be raised to 99% in some cases. Wet scrubbers can easily be fitted to existing power plants, provided the space is available. Wet scrubbing technology is technically complex. It has been likened to a chemical plant operating within a power station. For this reason it requires skilled staff to operate. Nevertheless is provides the best-proven method for removing up to 99% of the sulphur from a coal-fired power plant's flue gas stream.

Nitrogen oxides capture strategies

As with sulphur dioxide, it is possible to remove NO_x after they have been formed in the flue gas of a power plant. The process involves the injection of either ammonia gas or urea into the flue gas stream. The chemical reacts with the NO_x present, converting them into nitrogen and water.

If the ammonia or urea is injected into the hot flue gas stream, where the temperature is between 870°C and 1200°C, the reaction will occur spontaneously. This is called *selective non-catalytic reduction* or *SNCR*. At lower temperatures, however, a special metal catalyst is necessary to stimulate the reaction process. Where a catalyst is necessary, the process is called *selective catalytic reduction* or *SCR*.

SNCR will remove between 35% and 60% of the NO_x from the flue gas stream. The technology has been demonstrated in a number of power plants in the USA and Germany. Nevertheless some technical issues remain to be resolved. Ammonia contamination of ash and ammonia slip, the release of unreacted ammonia into the atmosphere, are both potential problems.

An SCR system operates at a lower temperature than an SNCR system. Typical flue gas temperatures are 340–380°C. At these temperatures the reaction between ammonia and NO_x must be accelerated by use of a solid catalytic surface. This is normally made from a vanadium–titanium material or a zeolite. The system is generally capable of removing 70–90% of the NO_x emissions from a flue gas stream.

There are two drawbacks to SCR. First, it can only be used with low sulphur coals (up to 1.5% of sulphur) and secondly it is expensive. The catalyst also requires changing every 3–5 years. Even with low sulphur coal, SCR can lead to the formation of sulphur trioxide which becomes highly corrosive on contact with water when it forms sulphuric acid. Strategies are being developed to capture sulphur trioxide.

Combined sulphur and nitrogen oxides removal

There are a number of processes under development which combine the capture of sulphur dioxide and nitrogen oxides in a single unit. Such systems aim to be more cost effective than a combination of an SCR system and a desulphurisation unit.

Processes that combine the two include an activated carbon-based process tested in Germany and a similar system with a proprietary absorbent material under test in the USA. Electron irradiation procedures have been developed and tested in Japan and the USA and an electrical discharge technique is also under test. These, and others, are still in the development or demonstration stage and are not yet proven for large-scale application.

Particulate removal

There are two principal systems for removing particulates from the flue gas of a coal-fired power station, electrostatic precipitators (ESPs) and fabric (baghouse) filters.

Invented by the American scientist Frederick Cottrell, the ESP is well established and the technology has been widely exploited. It utilises a system of plates and wires to apply a large voltage across the flue gas as it passes through the precipitator chamber. This causes an electrostatic charge to build up on the solid particles in the flue gas; as a result they are attracted to the oppositely charged plates of the ESP where they collect. Rapping the plates caused the deposits to fall to the bottom of the ESP where they are collected and removed. A new ESP will remove between 99.0% and 99.7% of the particulates from flue gas.

Bag filters, or baghouses, are tube-shaped filter bags through which the flue gas passes. Particles in the gas stream are trapped in the fabric of the bags from which they are removed using one of a variety of bag-cleaning procedures. These filters can be extremely effective, removing over 99% of particulate material. They are generally less cost effective than ESPs for collection efficiencies up to 99.5%. Above this, they are more cost effective. A system that combines a baghouse-style filtration system with an ESP is under development too. This aims to provide a cost-effective high removal-efficiency system, but has not yet been extensively demonstrated.

Mercury removal

Most coals contain a small amount of mercury and this can easily end up being discharged in the flue gas from a coal-fired power plant. In the USA the emission of mercury is to be regulated and this will necessitate the introduction of effective capture methodologies.

ESPs capture up to around 50% of the mercury emitted by a plant while bag filers can remove up to 99%. However the efficiency of collection depends on the precise chemical form of mercury being captured. In the near term it may be possible to improve the capture of mercury by injecting activated carbon particles into the flue gas stream before it reaches the particulate capturing system. Other technologies are also being developed but have yet to be tested on a significant scale.

Carbon dioxide

The primary combustible component of coal is carbon and when carbon burns completely in air, it is turned into carbon dioxide. Consequently, the combustion of coal produces large quantities of carbon dioxide. Currently the release of carbon dioxide is very broadly controlled by the terms of the United Nations Framework Convention for Climate Change (FCCC) as agreed in the Kyoto Protocol. This sets targets for carbon dioxide emissions that can probably be met by introducing efficiency measures and by the adoption of more renewable sources of generation. In the long term, however, it is looking certain that the capture and storage of carbon dioxide from the flue gases of power plants will become mandatory.

There are a number of methods of carbon dioxide capture under development. These can be broadly classified under chemical absorption, physical absorption and membrane separation. Chemical absorption involves using a chemical to capture and bind carbon dioxide. This chemical is then transferred to a processing plant where it treated to remove the carbon dioxide which is captured and stored. The chemical agent is then recycled through the power plant.

Physical absorption involves absorbing the carbon dioxide within a solid compound which is placed in its path in the flue gas stream. The solid is then treated, usually at low pressure, to remove the carbon dioxide, which is again stored. Membrane separation involves exploiting the properties of a special membrane which will allow carbon dioxide to pass through it but will not pass oxygen or nitrogen.

Separating the carbon dioxide from the flue gases of a coal-fired power plant solves only half the problem. The second half is to find somewhere to store it. In the short term there may be sufficient demand for carbon dioxide, industrially, to absorb some small-scale capture. One use that is being strongly touted is for enhanced oil recovery. This involves pumping carbon dioxide into oil reservoirs under pressure to force out additional oil. This procedure is already common practice and in many cases the carbon dioxide is specifically manufactured for the purpose. If power plant carbon dioxide could be sold for injection into underground oil reservoirs, this would significantly affect the economics of its capture, and could help make the technology more viable.

In the longer term, however, special strategies for storing carbon dioxide will be required. The gas could be stored in underground caverns, as natural gas has been for decades. Other options being explored involve storing it in porous-rock layers underground, or even at the bottom of oceans where the extreme pressures would keep it locked in a solid state. However storage may prove to be a greater technical challenge than capture.

Advanced coal-burning power plant technology

The traditional coal-fired power plant suffers two primary drawbacks. Firstly, its overall efficiency is limited and secondly it is a major source of pollution. There are strategies that can be applied to the traditional plant to dramatically reduce the levels of pollution produced. However there is little that can be done to increase its efficiency apart from raising the steam pressure and temperature. This requires expensive materials and may not be cost effective in the near future.

Alternative approaches to coal-plant design do exist. These allow plant emissions (particularly sulphur dioxide and nitrogen oxides in exhaust gas) to be controlled more simply and effectively. They may also offer some improvement in conversion efficiency. The most important of these technologies are fluidised bed combustion (FBC) and integrated-gasification combined cycle (IGCC).

Fluidised-bed combustion

If a layer of sand, of finely ground coal, or of another fine solid material is placed in a container and high-pressure air is blown through it from below, the particles, provided they are small enough, become entrained in the air and form a floating, or fluidised, bed of solid particles above the bottom of the container. This bed behaves like a fluid in which the constituent particles constantly move to and fro and collide with one another. As a type of reactor, this offers some significant advantages.

The fluidised bed was used first in the process industries to enhance the efficiency of chemical reactions between solids by simulating conditions of a liquid-phase reaction. Only later was its application for power generation recognised. Its use is now widespread, and the fluidised bed can burn a variety of coals as well as other poorer fuels such as coal-cleaning waste, petroleum coke, wood and other biomass.

A fluidised bed used for power generation contains only around 5% coal or fuel. The remainder of the bed is primarily an inert material such as ash or sand. The temperature in a fluidised bed is around 950°C, significantly lower than the temperature in the heart of a pulverised-coal boiler. This low temperature helps minimise the production of NO_x. A reactant such as limestone can also be added to the bed to capture sulphur, reducing the amount of sulphur dioxide released into the exhaust gas. One further advantage of the fluidised bed is that boiler pipes can be immersed in the bed itself, allowing extremely efficient heat capture (but also exposing the pipes to potentially high levels of erosion).

There are several designs for fluidised-bed power plants. The simplest is called a *bubbling-bed plant*. This, and a second, more complex plant called a *circulating fluidised bed* can both operate an atmospheric pressure. The circulating bed can remove 90–95% of the sulphur from the coal while the

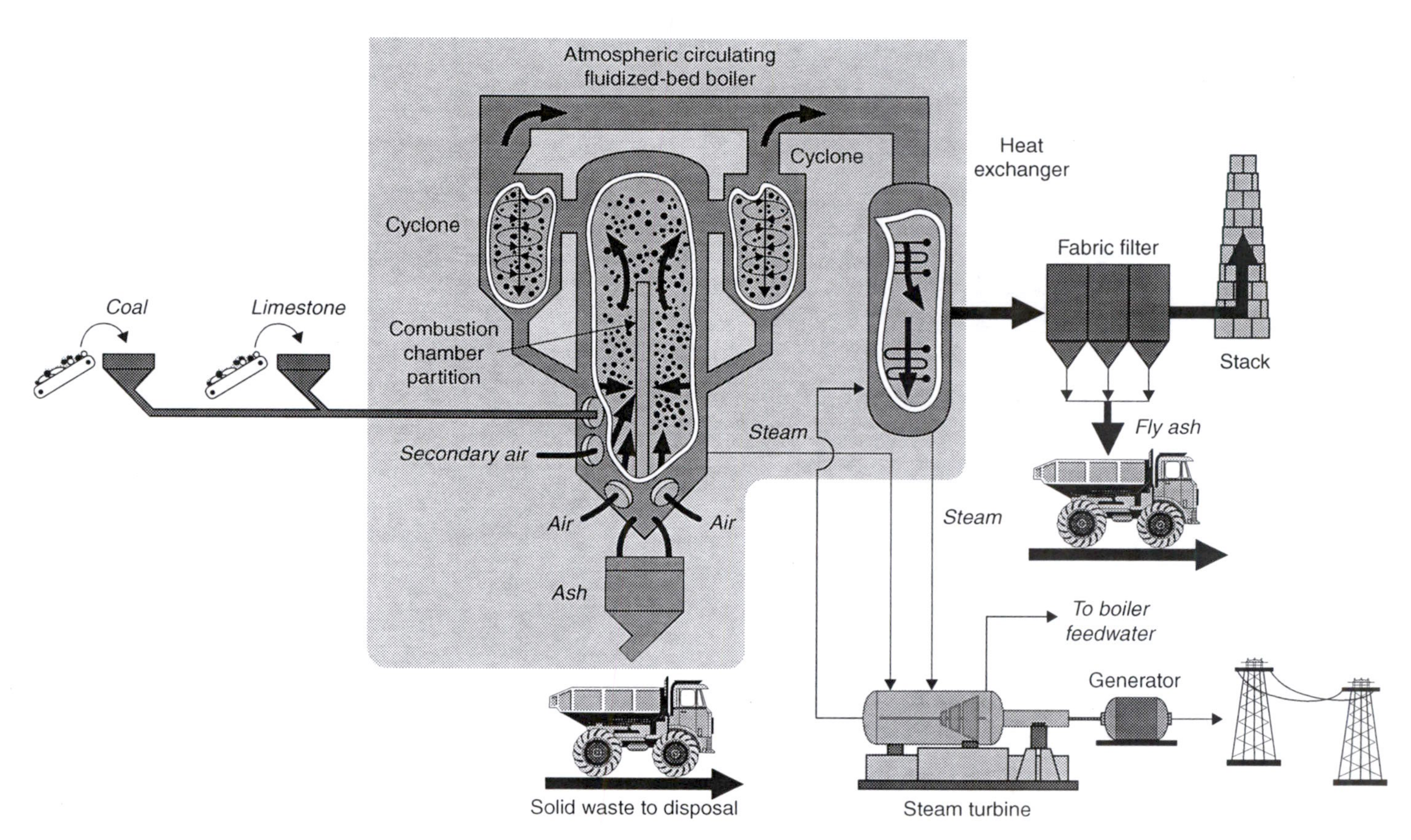

Figure 3.4 *Flow diagram for a circulating fluidised-bed power plant.* Source: *Tri-State Generation and Transmission Association, Inc.*

bubbling bed can achieve between 70% and 90% removal. Maximum energy conversion efficiency is 43%, similar to that of a traditional pulverised-coal plant. However such high efficiencies can only be achieved with larger plants that can employ larger, and generally more efficient, steam turbines under optimum steam conditions.

A third type of fluidised-bed design, called the *pressurised fluidised bed*, was developed in the late 1980s and the first demonstration plants employing this technology were constructed in the mid-1990s. The pressurised bed is like a bubbling bed, but operated at a pressure of between 5 and 20 bar (1 bar is equivalent to atmospheric pressure).

Operating the plant under pressure allows some additional energy to be captured by venting the exhaust gases through a gas turbine. This provides a higher efficiency (currently up to around 45–46%) while maintaining the good emission performance of the atmospheric pressure fluidised bed. The largest pressurised fluidised-bed plant in operation is a 360 MW unit in Japan.

Atmospheric fluidised-bed power plants with boiler capacities of over 400 MW are commercially available. These can provide supercritical steam to gain the best efficiency. The technology is still under active development, with the prospect of more efficient capture of pollutants coupled with an efficiency of around 50% within the next 10–15 years.

A standard fluidised-bed power plant can meet the emission-control requirements in many regions of the world without further emission-control measures. However in regions with the most stringent regulations capture technologies are required. These are likely to include NO_x, sulphur oxides (SO_x) and particulate capture measures. The techniques employed to provide additional emission control are the same as those used in a conventional coal-fired power station.

Integrated-gasification combined cycle

The second type of advanced coal-burning plant, the IGCC plant, is based around the gasification of coal. Coal gasification is an old technology. It was widely used to produce town gas for industrial and domestic use in the USA and Europe until natural gas became readily available.

Modern gasifiers convert coal into a mixture of hydrogen and carbon monoxide, both of which are combustible. Gasification normally takes place by heating the coal with a mixture of steam and oxygen (or, in some cases, air). This can be carried out in a fixed bed, a fluidised bed or an entrained flow gasifier.

The process that takes place in the gasifier is a partial combustion of the coal. Consequently it generates a considerable amount of heat. This heat can be used to generate steam to drive a steam turbine.

The gas produced, meanwhile, is cleaned and can be burned in a gas turbine to produce further electricity. Heat from the exhaust of the gas

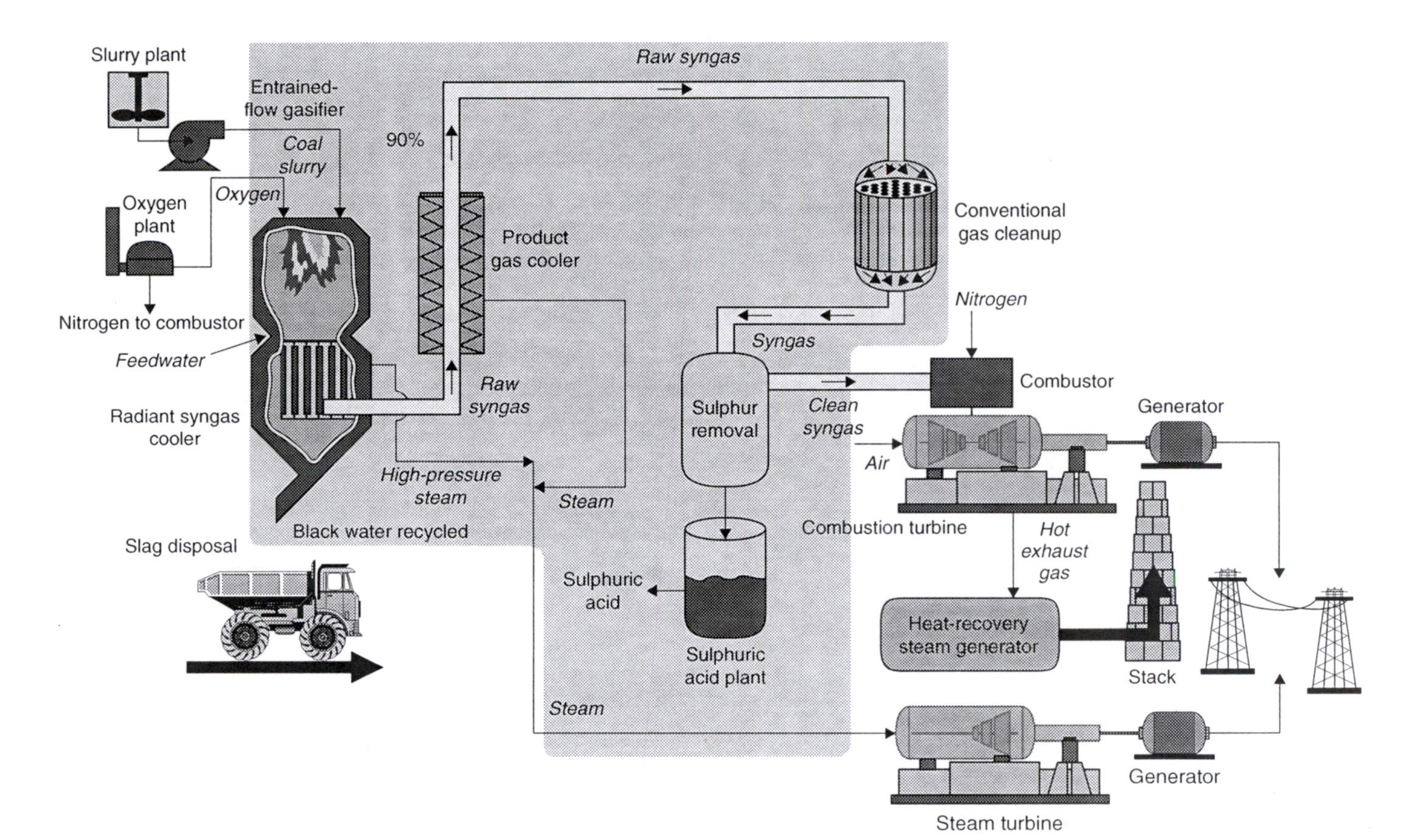

Figure 3.5 *Flow diagram of an IGCC plant*. Source: *Tampa Bay Electric Company*

turbine is used to raise additional steam for power generation. This is the basis of the IGCC plant.

An IGCC power plant can achieve an efficiency of 45%. In addition it can remove 99% of the sulphur from the coal and reduce the emissions of NO_x to below 50 ppm. Several demonstration projects were built in the mid- and late 1990s, with unit sizes up to around 110 MW. Since then a small number plants have entered commercial operation. Three are in Italy where they burn high sulphur residues at oil refineries. Like these Italian plants, most applications are in niche markets such as the oil refinery or chemical industries. The technology has yet to make an impact in the main power generation market.

Further development is required to enable gasification to realise its full potential. This will include effective technologies for cleaning the hot exhaust gas before it enters the gas turbine stage of the IGCC plant. Hot gas cleanup will allow an IGCC plant to operate at optimum efficiency.

One aspect of gasification technology which has attracted recent attention is its ability to produce gaseous hydrogen. If an energy economy based on hydrogen ever evolves, then coal gasification could provide one source of the fuel.

Another area that could prove attractive is underground gasification. This involves the controlled burning of coal in the seams underground where it is found. Air is injected through a borehole into the seam and the gasification product is extracted from a second borehole. Underground gasification avoids many of the pollution problems associated with coal combustion while requiring little advanced technology. However the technique is nowhere near commercial application.

Environmental effects of coal combustion

Uncontrolled coal combustion is a filthy process. It releases a catalogue of unpleasant solids and gases that are potential environmental contaminants. Fortunately various strategies exist, as outlined above, to contain most of these potential pollutants. But even with the most effective pollution-control systems, some environmentally detrimental materials are released.

In many countries the release of the most obvious contaminants such as sulphur dioxide, nitrogen oxides and particulates are controlled by environmental regulations. These may either specify the maximum concentration of pollutant permitted in the flue gas exiting a power station or they may set an upper limit for the total amount of each material that can be released into the environment. In either case, power plant operators will aim to keep within these limits, but only just. Some typical regulation levels are shown in Table 3.2.

Without regulation, all these pollutants cause severe environmental damage. SO_x and NO_x are responsible for acid rain which has caused widespread damage to forests and lakes throughout the world. Acid rain

Table 3.2 *Annual average emission standards (mg/m³)*

	Sulphur dioxide	*Nitrogen oxides*	*Dust*
European Union	400	500	50
Japan	170–860	410	50
USA	1500	700	60
World Bank (recommended)	2000	750	50

Source: World Bank recommendations, European Union 2001 limits for large power plants, World Bank Technical Paper No. 286.

can also cause damage to stonework as may be seen from the Taj Mahal to the centre of London. NO_x help cause smog, as do the ash particles released from a power plant. All these pollutants can cause health problems when inhaled and their effects are compounded when significant amounts of trace metals such as mercury, lead or cadmium are also released.

More serious perhaps, certainly potentially more far-reaching in its effects, is the threat of global warming as a result of the release of excessive quantities of carbon dioxide into the atmosphere. Carbon dioxide is one of a number of gases (methane and SO_2 are others) which have been implicated in the greenhouse effect. These gases allow heat from the sun to reach the surface of the earth but then prevent it from being re-radiated into space. The effect is to raise the average global temperature (see Chapter 2).

So far the control of carbon dioxide emissions has been left to the broad targets set by the Kyoto agreement. These aim, as a first step, to return the rate of emission of carbon dioxide to levels around 6–8% below those seen in 1990 by between 2008 and 2012. (These targets only apply to the industrialised nations. Developing nations are broadly exempt though they will be embraced by the treaty at a later date.) This target is being achieved, without much effort in some cases, by a switch to the combustion of natural gas rather than coal or oil and by the steady increase in the efficiency of modern power plants. A downturn in the global economy at the end of the twentieth century has also had a beneficial effect.

This is a short-term measure. In the long term it will be necessary to legislate to introduce the technology to capture and store carbon dioxide. Such legislation will have a significant effect on the economics of coal-fired power plants. However the time scale for the introduction of such measures is far from clear and the resistance from the interested parties is intense.

Financial risks associated with coal-fired power generation

The basic technology involved in burning coal to generate electricity using a steam turbine is more than one-century old and the optimum power plant configuration, materials and operating conditions are extensively documented. From the standpoint of the simple boiler and steam turbine plant burning pulverised coal, behaviour is predictable and the technical risks minimal.

Power plant performance will depend on the exact type of coal to be burned and the nature of the fuel must be taken into account. This will generally mean designing a plant around a particular type of coal, often from a single source. Here factors such as the extent and reliability of the supply must be carefully weighed too. Later adaptation to a different coal from a different source is possible but it will affect generating costs.

Also critical from the design standpoint are the environmental regulations that exist when the plant is to come into service, and any possible changes to these regulations in the future. New plants constructed in most developed countries will require extensive emission-control systems. In some developing countries the existing regulations are often less stringent but this is a position that is likely to change. It would be wise to at least make provision for the addition of flue gas treatment systems in any new plant.

The more advanced coal-burning technologies such as FBC and IGCC are relatively recent innovations. Atmospheric FBC systems have been extensively demonstrated in power generation applications and their reliability is generally proven. Risks should be minimal and these plants offer the added flexibility of being able to burn different coals with ease. This takes some of the risk out of fuel supply.

Pressurised FBC plants and IGCC plants are still in a relatively early commercial stage. Long-term reliability has to be established and some component development remains. Commercial implementation of these technologies must be considered to carry an additional risk to that attaching to the more established technologies for burning coal.

Aside from the technology, the second source of risk associated with coal-fired power plants relates to fuel supply and cost. Both have remained relatively stable in the last 20 years, though there have been some price peaks (see below). Where coal is abundant, and particularly where it faces competition from natural gas, this situation appears likely to continue for the immediate future. Over the lifetime of a new coal-fired plant, prices will show an increase but with several major sources around the globe, no dramatic move is likely. This situation should allow for firm planning of fuel costs.

The situation may not prove so straightforward where a plant is built to exploit a local source of coal. Fuel costs under these circumstances are likely to be lower but that advantage could be counterbalanced by less security of supply and less price stability. In order to ensure that the project is to remain viable over its lifetime, a secure fuel supply agreement is vital. And in a country where the fuel delivery might be unreliable, some form of guarantee should be sought.

The cost of coal-fired electricity generation

A decision to build a coal-fired power station will depend on many factors such as fuel availability, the environmental hazards attached to the project, and the cost of alternative methods of electricity generation. A coal-fired plant will generally be built for base-load duty, though more modern plants and technologies do allow some load following without grave economic penalties. In general, however, economic viability must be established in comparison with other base-load generating technologies such as hydropower, nuclear power and gas-fired combined cycle power stations.

As with all fossil-fuel-fired technologies, the cost of electricity depends on both the cost to build the generating plant and the cost of the fuel. Coal-fired power stations tend to cost more than gas turbine power plants, but coal is usually cheaper than gas. Table 3.3 collects estimates from three sources for the capital costs of different coal-fired power station technologies. The cost of a new conventional plant with emission-control systems will vary depending on the efficiency of the capture. The estimates in Table 3.3 assume nitrogen oxides, sulphur dioxide and particulates are all being controlled to meet US regulation levels. With less stringent restrictions, capital costs could be reduced.

The table shows that a conventional coal-fired power plant costs less than an atmospheric fluidised-bed power plant. The cost comparison with the pressurised fluidised-bed plant is more difficult, but when the efficiency of the pressurised plant is taken into account, it could win. IGCC, too, is more expensive than a conventional plant, but again the additional efficiency will have a significant effect on long-term levelised costs for electricity.

The Energy Information Administration (EIA) has published estimates for annual operating and maintenance (O&M) costs for coal-fired power plants. Its most recent estimates put the fixed O&M costs for a pulverised-coal plant at US$22.5/kW and the variable O&M costs at USmills3.25/kWh. For an IGCC plant the fixed O&M costs are US$24.2/kW and variable costs are USmills1.87/kWh.

Table 3.3 *Capital costs ($/kW) of coal-fired power plants*

	CEED	*World Bank*	*EIA*
Conventional plant with emission control	1400	–	1079
Atmospheric fluidised bed	1500–1800	1300–1600	–
Pressurised fluidised bed	1250–1500	1200–1500	–
IGCC	1500–1800	1500–1800	1200–1800

Source: Center for Energy and Economic Development (CEED), World Bank Technical Paper No. 286, US EIA.

Most countries in both the developed and the developing world have the capacity to manufacture pulverised-coal-fired boilers for power generation applications. Steam turbine manufacture, too, is widespread although the most efficient machines still come from established manufacturers in the USA, Europe and Japan. With indigenous capability, the need for foreign exchange to fund construction is reduced, making conventional coal-fired capacity attractive. The advanced coal-fired systems generally require a higher level of technological expertise to manufacture. As a result, core components of these power plants will often have to be imported. This will make such plants less attractive, particularly in the developing countries where funds are scarce.

The bottom line for any coal-fired project, however, is the cost of the fuel. Where the fuel is available locally, as it is in many parts of the world, it will always prove attractive and will frequently provide the cheapest source of electricity.

In countries such as the USA and China, where coal is plentiful, coal-fired power generation will generally be competitive,[4] even when the cost of pollution-control technologies are taken into account. But factors affecting the cost of coal are critical. In particular, transportation can seriously affect the economics of generation. Coal is expensive to haul and the shorter the distance the better. Where it must be transported, cheap bulk transport it is important. In the USA, the cheapest form of coal transportation is by barge, followed by truck and then rail.

Countries without their own resources have to rely on imported coal. World coal prices began to climb in 1994 and peaked in the third quarter of 1995 at around $45/tonne (spot price for coal in north European ports). By the middle of 1997, they had fallen back somewhat, to around $40/tonne and in 2000 they were close to $33/tonne. Estimates suggest that a price of around $45–50/tonne is necessary to fund the development of new mines. But with relatively few buyers and a large number of suppliers, it seems unlikely that there will be a major change in the cost of coal.

End notes

1 World Energy Council, Survey of Energy Resources, 2001.

2 Costs for coal and coal-cleaning technologies used in this chapter are almost exclusively taken from E. Stratos Tavoulareas and Jean-Pierre Charpentier, World Bank Technical Paper No. 286: Clean Coal Technologies for Developing Countries published in July 1995. Though this report is now 8-year old, the figures still remain relevant. This is also the only available contemporary source which examines the whole range of technologies and compares them on the same basis.

3 This type of boiler is called a *watertube boiler* because the water circulates in tubes within the combustion gases. There is a simpler design called a

firetube boiler in which the combustion gases are contained within tubes that pass through water. This type of design is used for small and domestic boilers, but is not now used in power plants.

4 It is worth noting that though new coal-fired power plants were scare in the USA during the 1990s, when there was a strong move towards gas-fired combined cycle plants, the beginning of this century has seen a renaissance in the coal-fired plant.

4 Gas turbines and combined cycle power plants

The gas turbine has seen a recent and meteoric rise in popularity within the power generation industry. Until the end of the 1960s gas turbines were almost exclusively the preserve of the aviation industry. During the 1970s and 1980s they started to find favour as standby and peak power units because of their facility for rapid start-up. It was during the 1990s, however, that they became established, so that by the end of the twentieth century the gas turbine had become one of the most widely used prime movers for new power generation applications – both base load and demand following – virtually everywhere. It has been suggested that gas turbines could account, for example, for 90% of new capacity in the USA in the next few years.

A number of factors contributed to this change of fashion. Deregulation of gas supplies, particularly in the Europe and the USA, and the rapid expansion of natural gas networks have increased the availability of gas while conspiring to keep prices of natural gas low. More and more stringent emission-control regulations have pushed up the cost of coal-fired power plants making relatively pollutant-free natural gas look more attractive. Power sector deregulation has also contributed, by attracting a new type of generating company seeking quick returns. Gas-turbine-based power stations can be built and commissioned extremely rapidly because they are based around standardised and often packaged units and the capital cost of gas turbines has fallen steadily, making then economically attractive to these companies.

The most potent factor, however, has been the development of the combined cycle power plant. This configuration, which combines gas and steam turbines in a single power station, can provide a cheap, high-capacity, high-efficiency power generation unit with low environmental emissions. With net conversion efficiencies of the largest plants now around 50%, and with manufacturers claiming potential efficiencies of 55% or more in plants incorporating their latest machines, the combined cycle plant offers power generating companies a product that seems to promise the best of economic and environmental performance that technology can currently offer.

This unrestrained popularity has occasionally led power generating companies into difficulties. In the UK, for example, there was a significant move towards gas-fired combined cycle power plants during the 1990s. New market regulations introduced at the end of the decade led to a marked fall in electricity prices and may combined cycle plants could no longer generate power economically.

This inconvenient conspiracy of economic factors highlights the main factors in the gas turbine for power generation equation. Gas turbines are cheap but the fuel they burn, normally natural gas, is relatively expensive. The economics of gas-based generation is therefore extremely sensitive to both electricity and to gas prices. Gas turbines can burn other fuels, distillate or coal-bed methane for example. However the modern boom is based on natural gas and it is upon this that their continued progress will rest.

Natural gas

The switch from coal- and oil-fired power plants to natural gas-fired plants has become a global phenomenon. This is reflected in gas production and consumption statistics. World Energy Council Figures[1] indicate that the production of natural gas increased by 4.1% between 1996 and 1999. In China gas use increased by 10.9% in 1999 and in the Asia-Pacific region the increase was 6.5%. Africa's consumption increased by 9.1%.

Globally the USA was the largest consumer of natural gas in 2001 according to the US Energy Information Administration (EIA)[2] followed by Russia, Germany, the UK and Canada. Russia and the USA, meanwhile, were the main producers, accounting between them for 44% of annual production in 2001. They were followed by Canada, the UK and Algeria.

In Europe gas usage is expected to increase dramatically during the next two decades. According to Eurogas[3] consumption will rise from 332 million tonnes of oil equivalent (mtoe) in 2000 to 471 mtoe in 2020, a rise of 42%. Europe's principal users in 2000 were the UK, Germany, Italy, France and the Netherlands. Of these only the UK and the Netherlands produce significant quantities of gas. The other countries import most of the gas they consume.

Of course not all this gas is burned in power stations, but a significant proportion of it is. In the USA, for example, power generation accounted for around 20% of natural gas in 2001. As has already been noted, the driving forces behind the increasing popularity of the fuel within the power industry are economic – gas turbines are cheap and can be deployed rapidly – and environmental. Natural gas produces lower levels of atmospheric pollution that either coal or oil when it is burned. This includes sulphur dioxide, nitrogen oxides (NO_x), hydrocarbon particulates and carbon dioxide. Thus it is easier to meet emission regulations with a gas-fired power plant than it is with a plant burning either coal or oil.

The gas industry is keen to promote the idea of gas as a clean fuel but critics would argue that its use is at best a stopgap. A sustainable energy future must rely on renewable sources of energy and gas is not renewable. More importantly, the supply of gas available in the world is limited.

Table 4.1 *Proved recoverable natural gas reserves*

	Reserve (billion m^3)	*Estimated reserve life (years)*
Africa	11,400	69
North America	7943	9
South America	6299	63
Asia	17,106	52
Europe	53,552*	58
Middle East	53,263	>100
Oceana	1939	46
Total	151,502	58

*The Russian Federation contributes 47,730 billion m^3 to this total.
Source: World Energy Council.

As Table 4.1 shows, current proven reserves are expected to last for around 60 years at current levels of consumption.

Table 4.1 lists the estimated recoverable natural gas reserves from different regions of the world, based on figures collated by the World Energy Council for its 2001 Survey of Energy Resources. As these figures illustrate, Europe and the Middle East have the largest proven recoverable reserves. (*Note*: however, that most of the European reserves are located in the Russian Federation.)

North America and Western Europe are taxing their known reserves most heavily. At 1999 rates of gas production, proved reserves in the USA would be exhausted within 9 years. However the estimated reserves remain enormous so this is no immediate cause for concern. In Western Europe, the Netherlands and Norway both have extensive reserves remaining. Elsewhere proven reserves are in a similar or worse situation to that in the USA. Indeed Western Europe is having to rely increasingly on imports, primarily from Russia and Algeria, to maintain its supplies of gas. From an energy security perspective, this could become a dangerous situation in the future.

Natural gas costs

The use of natural gas to generate electricity depends crucially on the cost of the gas. Natural gas is a more costly fuel than coal, the other major fossil fuel used for power generation. However the capital cost of a coal-fired power plant is significantly higher than that of a gas-fired power station. Hence the total fuel bill over the lifetime of each plant determines whether coal or gas can produce the cheapest electricity.

Utility gas prices are often closely linked to the price of oil, though deregulation of the gas industry has weakened the link in some countries

Table 4.2 *Global gas prices for power generation ($/GJ)*

	1997	*1998*	*1999*	*2000*	*2001*	*2002*
Finland	3.06	2.87	2.58	2.70	2.61	2.61
Germany	3.78	3.51	3.35	3.66	–	–
Taiwan	6.10	5.23	4.83	5.88	5.86	–
UK	2.94	3.01	2.75	2.51	2.65	1.94
USA	2.63	2.25	2.44	4.11	4.42	3.42

Source: US Energy Information Administration.

such as the UK. One reason for this link is that many gas-fired power plants can easily be fired with oil and would switch to oil if natural gas became more expensive. This fixes an upper limit on the cost of natural gas. (It is worth noting, however, that while some gas-fired steam plants can burn residual oil, gas turbines require distillate which is more expensive. Even so, most gas turbine plants are designed for dual fuel use, that is gas or oil.)

Table 4.2 collects annual prices of gas for power generation from a handful of countries between 1997 and 2002. These give a broad indication of how costs vary across globe. The Finnish prices in the table are remarkably stable over the 6-year period, whereas in the UK, princes fluctuated much more. However the USA showed the largest range of prices, with the cost of gas for power generation soaring in 2000 and 2001. Such volatility can play havoc with power generation economics.

Where gas supplies are limited or non-existent the possibility exists to import liquefied natural gas (LNG). LNG costs more than piped gas when the cost of liquefaction, transportation and regasification are taken into account. This is illustrated in Table 4.2 with the gas prices for Taiwan which are consistently the highest quoted. Even as such a high price, LNG has proved attractive to countries like Japan, Taiwan and South Korea. In 1999 25% of exported natural gas was in the form of LNG.[4] Of this 75% was transported to the Asia-Pacific region.

Gas turbine technology

A gas turbine is a machine which harnesses the energy contained within a fluid – either kinetic energy of motion or the potential energy of a gas under pressure – to generate rotary motion. In the case of a gas turbine this fluid is usually, though not necessarily, air. The earliest man-made device for harnessing the energy in moving air was the windmill, described by Hero of Alexandria in the first century AD.

The early windmill was a near relative of today's wind turbine. Closer in concept to the gas turbine was the smokejack, developed in the middle

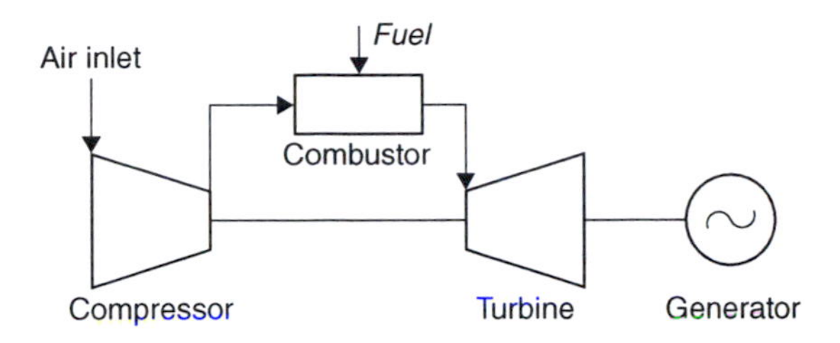

Figure 4.1 *Block diagram of a gas turbine for power generation*

of the second millennium AD. As described in the seventeenth century by John Wilkins, later Bishop of Chester, the smokejack used hot air rising through a chimney to move windmill vanes and drive a shaft which could be used to rotate a spit for roasting meat.

This principle of harnessing moving air to create rotary motion for driving machinery was developed further during the industrial revolution. following this principle, the nineteenth century saw a number of predecessors to the gas turbine. These used some form of compressor to generate a flow of pressurised air which was fed into a turbine. In these machines the compressor was usually separate from the turbine.

The direct ancestor of the modern gas turbine was first outlined in a patent granted to German engineer F. Stolze in 1872. In Stolze's design, as in that of all modern gas turbines, an axial compressor was used to generate a flow of pressurised air. This air was then mixed with fuel and ignited, creating a flow of hot, high-pressure gas which was fed into a turbine. Crucially the compressor and the turbine were mounted on the same shaft.

Whereas a gas turbine supplied with pressurised gas from a separate compressor must inevitably rotate provided it has been designed correctly, the arrangement patented by Stolze need not necessarily do so. This is because the energy to operate the compressor which provides the pressurised air to drive the turbine is produced by the turbine itself. Thus unless the turbine can generate more work than is required to turn the compressor – the energy for this being provided by the combustion of fuel which produces the hot gas flow to drive the turbine – the machine will not function. This, in turn demands extremely efficient compressors and turbines. Both need to operate at an efficiency of around 80%. In addition the turbine must be able to accommodate very hot inlet gases in order to derive sufficient energy from the expanding gas flow. Only if these conditions are met will the turbine operate in a continuous fashion.

The turbine system described by Stolze, although envisaging virtually all the features of a modern gas turbine, was not capable of sustained operation because the machinery to achieve it had not yet been developed. The first machine, which could run in a sustained fashion, was built in Paris in 1903. This, though, did not have a rotary compressor on the same axis as the turbine. That honour fell to a machine built by Aegidus Elling in Norway and operated later in 1903. In Elling's machine the inlet gas temperature was 400°C.

Development of the gas turbine continued through the early years of the twentieth century, the aim remaining to generate either compressed air, rotary motion or both for industrial use. Then, during the 1930s, the potential of the gas turbine to provide the motive force to flight was recognised and aircraft with jet engines based on the gas turbine were developed in Germany, in Great Britain and in the USA. These led, in turn, to the modern aircraft engines that power the world's airline fleets.

During the late 1970s and early 1980s gas turbines began to find a limited application in power generation because of their ability to start up rapidly. This made them valuable as reserve capacity, brought into service only when grid demand came close to available capacity. These units were based on the aeroengines from which they were derived but by the late 1980s larger, heavy gas turbines were under development. These were intended solely for power generation.

Modern gas turbine design

The key to gas turbine operation is efficiency; efficiency both of the compressor and of the turbine. Each must be adequately efficient to overcome the natural barrier to sustained operation. Beyond that, the more efficiency the machinery, the more effective it becomes.

High efficiency of operation is also one of the key factors in the popularity of modern gas turbines for power generation. The more efficient a gas turbine, the more electricity it can produce from a given quantity of fuel. But efficiency is also important from an environmental perspective too. The higher the efficiency of a fossil-fuelled power plant, the smaller the quantities of atmospheric pollutants it produces for each unit of electricity. In this regard, gas turbines score highly.

Efficiency is equally important in the aero industry. But turbines developed for the aviation applications must also be light and extremely reliable. For power generation weight is not a significant factor but cost is. As a result the development paths for the two types of turbine have diverged.

As already outlined, the first designs for gas turbines utilised separate compressors and turbines. Stolze's design simplified this by putting the compressor and turbine on a single shaft so that the power generated by the turbine would drive the compressor as well as producing mechanical output.

Modern gas turbines for power generation applications generally utilise axial compressors with several stages of blades (like a series of windmills, but working in reverse) to compress air drawn in from the atmosphere to perhaps 15–19 times atmospheric pressure. These compressors have efficiencies of around 87%. A modern unit might have 10–12 sets of compressor blades (stages).

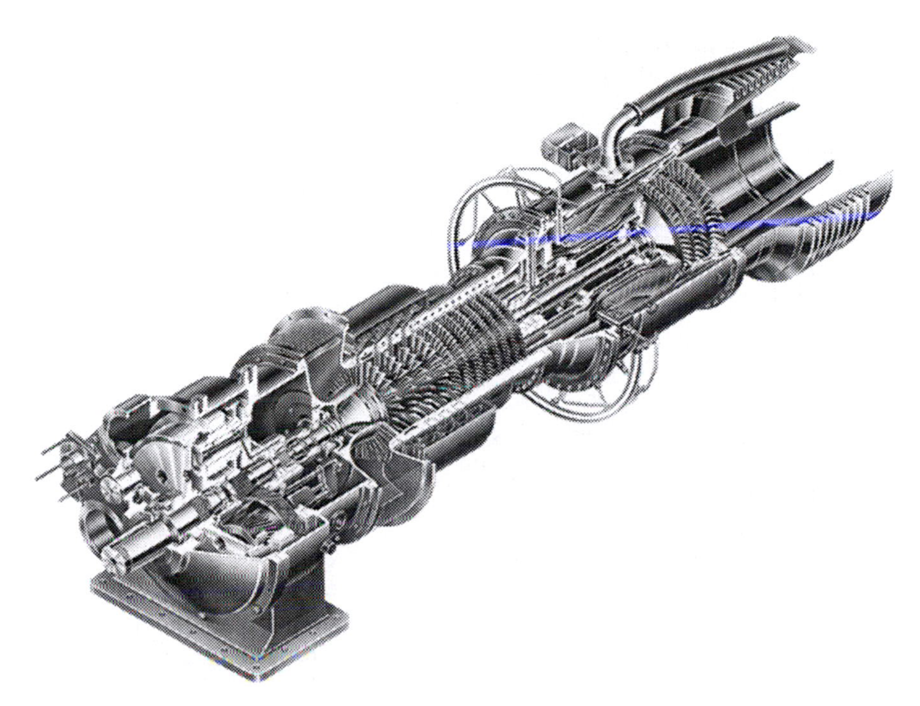

Figure 4.2 *Cross section (photograph) of a gas turbine.* Source: *Courtesy of Solar Turbines Incorporated*

High-pressure air from the compressor then enters a combustion chamber where it is mixed with fuel and ignited, increasing the temperature of the air to as much as 1400°C, or higher in some of the latest machines to appear. The gas turbine combustion chamber is specially designed to produce the minimum quantity of NO_x. This NO_x is produced at high temperature by a reaction between oxygen and nitrogen in air, but this can be controlled by controlling the combustion process so that all the oxygen is used during combustion, leaving none to react with nitrogen.

Combustion chambers come in a variety of designs and dispositions. In some gas turbines they are kept separate from the turbine body. Others designs position them within the body, between compressor and turbine stages, while in others there are multiple combustion chambers arranged annularly around the body of the turbine.

The hot air exiting the combustion chamber must have its temperature carefully controlled so that it cannot damage the first stage of the turbine. It is important, however, that the temperature should be as high as possible for the best possible efficiency, and as materials have improved, so inlet temperatures have risen. In 1967, there were typically around 900°C, reaching 1100°C in the 1970s. By 2000 materials could cope with an inlet temperature of 1425°C.[5]

The turbine stage of a modern gas turbine will normally comprise three to five stages of blades (windmills operating as windmills in this case) operating with an efficiency of around 89%. Some designs have both compressor and turbine blades mounted rigidly onto the same shaft. In others there are two concentric shafts, one carrying the compressor blades and the first one or two turbine stages. These turbine stages power the compressor while the latter stages, on a second shaft, are attached to a generator and produce power.

Small gas turbines, with outputs of 35–45 MW, can achieve energy conversion efficiencies of up to 38% in power generation applications. Larger gas turbines usually for base-load combined cycle power plants, have traditionally shown slightly lower efficiencies but new, optimised designs have pushed efficiencies as high as 38.7% for modern large turbine designs.[6] These units can have outputs of 265 MW.

Since the maximum efficiency of a gas turbine depends on the temperature of the compressed air as it enters the turbine from the combustion chamber, much modern development has focussed on new and better materials that can withstand higher and higher temperatures. This has included such sophisticated materials as single crystals for first stage turbine blades. Ceramics are also being used as an alternative to metal.

Other factors can affect turbine performance. Intake air must be carefully filtered to prevent the entry of particles which could damage blades at the high velocities which are reached inside compressor and turbine. Injecting water into the compressor with air can improve efficiency. And with the latest high-temperature turbines, some form of blade cooling is often required. Thus gas turbines are perhaps the most sophisticated machines in regular use within the power generation industry and require very specialised design and manufacturing facilities.

Advanced gas turbine design

A gas turbine aeroengine must remain light and compact so it is not possible to add to it significantly in order to improve its performance. The stationary turbine for power generation does not suffer this restriction. Taking advantage of this greater freedom, engineers have explored a number of strategies that can be applied to stationary gas turbines in order to provide significant performance enhancements.

Reheating

In large steam-turbine-based power plants it is traditional to split the turbine into separate sections, one handling high-pressure steam, one handling medium-pressure steam and a third handling low-pressure steam. By splitting the turbine in this way, efficiency gains can be made through matching the individual turbine sections to operate under a narrower range

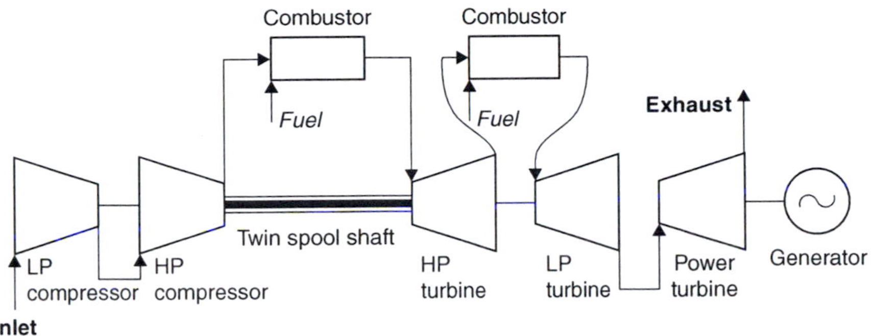

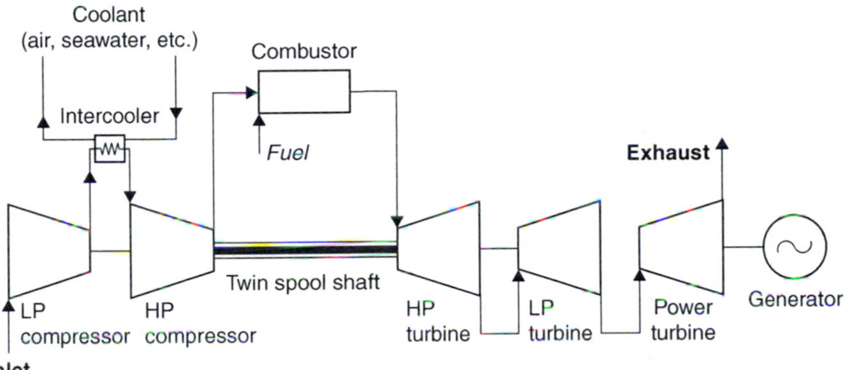

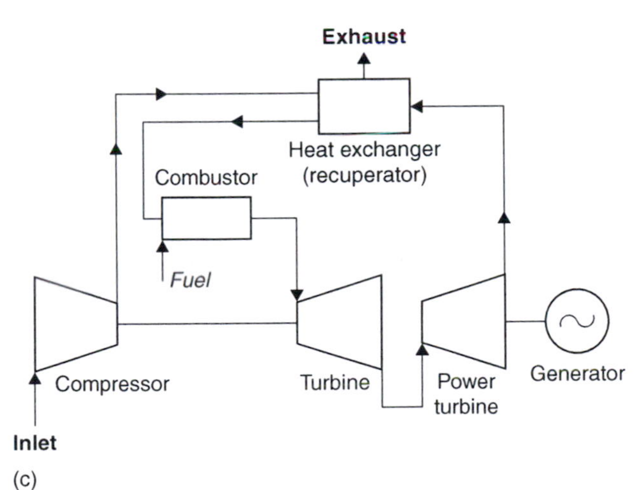

Figure 4.3 *Block diagram showing advanced gas turbine cycles: (a) reheating, (b) intercooling and (c) recuperation. LP: low pressure; HP: high pressure*

of steam pressures. Further, once the turbine has been split into separate sections, additional efficiency gains can be made by reheating the steam when it exists the high-pressure turbine (where it will have cooled) and before it enters the medium-pressure turbine. This is a common feature of the steam turbines used in coal-fired power plants.

A gas turbine can also be split in a similar way, though normally only two separate sections, called *spools*, are used. But again, once the turbine has been split into sections, it is possible to introduce a second combustion stage to reheat the air between the higher-pressure and the low-pressure section of the turbine. Using reheating makes the turbine more efficient, just as in the case of the steam turbine.

Reheat is already making an appearance in gas-turbine-based power plants. A 1000 MW plant in Monterrey in Mexico uses four gas turbines in which the hot gas is passed through a second combustor after the first stage of turbine blades before passing through the remaining four sets of blades.

Intercooling

It is possible to go a stage further with a gas turbine, by splitting the compressor into two sections: a low-pressure compressor section and a high-pressure compression section. And like the reheating of the air between the two sections of the turbine, it is possible to improve efficiency by cooling the air between the two sections of the compressor. (Compressing air tends to heat it and hot air occupies a larger volume. Cooling it reduces the volume so the compressor actually has less work to do.) This is called *intercooling*.

Intercooling a high-performance aeroderivative gas turbine (that is, a gas turbine for power generation based directly on an aeroengine) will boost its efficiency by around 5%, double its power output and substantially reducing the cost per kilowatt of generating capacity.[7]

Mass injection

Yet another strategy for increasing the efficiency of an aeroderivative gas turbine is to inject water vapour into the compressed air before the gas turbine combustion chamber. This system, called the *humid air turbine cycle* (HAT cycle), has a history dating back to the 1930s but it was only during the 1980s that an effective way of building such a turbine was devised.

The HAT cycle works because it requires less work from the compressor to deliver the same mass of gas into the turbine. The mass of water added to the compressed air tips the balance. It has been estimated that a 11 MW cascaded HAT cycle (CHAT cycle) unit incorporating humid air, intercooling and reheat could achieve an efficiency of 44.5%.[8] More striking still, a 300 MW CHAT turbine system would have an estimated efficiency of 54.7% and could prove cheaper than a gas-turbine combined cycle plant.

One disadvantage of HAT and CHAT cycle power units is that they release a considerable amount of water vapour into the environment. In situations where water is scarce it may be necessary to recover the water from the exhaust gas.

Recuperation

A fourth strategy for improving the performance of a gas turbine is to use heat from the turbine exhaust to partially heat the compressed air from the compressor before it enters the combustion chamber. This process, referred to as recuperation, results in less fuel being needed to raise the air to the required turbine inlet temperature.

Effective recuperation systems have been under development for several years. At the end of 1997, the US company Solar Turbines introduced a 3.2 MW gas turbine for power generation applications with a claimed efficiency of 40.5% using recuperation. This unit was developed under the US Department of Energy (DOE) Advanced Turbine Systems (ATS) programme. Other companies involved in the ATS programme include Pratt and Whitney which is developing a high-efficiency small gas turbine and GE Power Systems and Siemens-Westinghouse, both of which are working on high efficiency, large base-load combined cycle units. These are expected to yield overall efficiencies of 60% combined with low emissions.

Distributed generation

One of the roles envisaged for highly efficient, small gas turbine power units is distributed generation. This refers to a power-supply system where small generating units are installed close to the source of demand.

Distributed generation is particularly attractive in situations where there are centres of electricity demand at the end of long transmission lines, distant from major central power stations. Installing a small generating unit close to such a demand centre both improves the stability of the overall transmission and distribution network and reduces the need to upgrade the transmission system.

There are a number of electricity generating technologies that are well suited to distributed generation. These include fuel cells, solar and wind power and small gas turbine power units.

Combined cycle power plants

A single gas turbine connected to a generator can generate electricity with a fuel-to-electricity conversion efficiency of perhaps 38% using the best of today's technology. New developments, such as those falling under the auspices of the US DOE's ATS programme aim to push the simple cycle efficiency as high as 41% without cycle adaptation, 43% with adaptation such as recuperation. This is still marginally lower than a modern coal-fired power plant can hope to achieve.

Part of the reason for this lower efficiency resides in the fact that the exhaust gas leaving the gas turbine is still extremely hot; that is, it still contains a significant amount of energy which has not been harnessed to generate electricity. There are a wide variety of applications in which this exhaust heat can be used to generate hot water or steam for use in some industrial process, or for heating purposes. This forms the basis of a gas turbine co-generation system, a topic which will be covered in a separate chapter.

There is a second strategy which can be employed. The exhaust heat can be captured in a steam boiler – normally called a *heat recovery steam generator* (HRSG) – where it generates steam which is used to drive a steam turbine and create additional electricity. This is the basis for the combined cycle power plant.

Combined cycle plants may employ one, or several gas turbines. Normally each gas turbine is equipped with its own waste-heat boiler designed to capture the exhaust heat as efficiently as possible. In a power plant with more than one gas turbine, each may have its own steam turbine, or the units may be grouped so that several gas turbines supply steam for a single steam turbine.

A combined cycle power plant can be constructed from already available components, but the most efficient plants will employ gas turbines, HRSGs and steam turbines that have been matched to one another. While turbines are manufactured and then shipped to power plants site, the HRSG is built at the site. Two types of HRSG are in common use, horizontal and vertical. In a horizontal HRSG the exhaust gas from the gas turbine

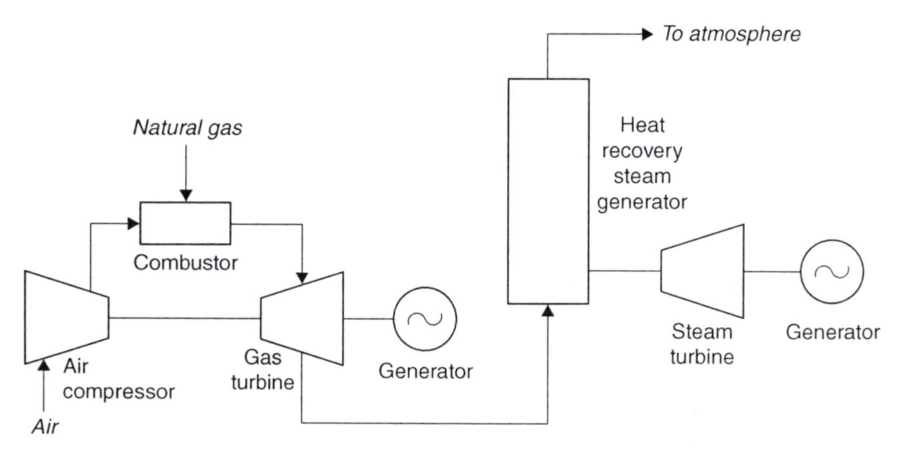

Figure 4.4 *A block diagram of a combined cycle power plant*

passes through it horizontally and the water/steam pipes which collect the heat are hung vertically in its path. The vertical HRSG reverses these arrangements. Vertical HRSGs are most popular in Europe where space for power plant development is restricted. Horizontal units are most popular in the USA.

Power plants based on the combined cycle configuration have become the workhorses of independent power producers all across the world. With individual heavy-frame gas turbines available in unit sizes up to 265 MW, such plants can be based on modules of around 300–400 MW. Actual power output can be increased by adding some additional heat generation within the HRSG, a procedure called *supplementary firing*.

Using the combined cycle configuration, power stations can be brought into service rapidly, with the gas turbine operating first in simple cycle mode, while the waste-heat recovery boilers and steam turbines are added later. Generating capacity can easily be increased incrementally too, by adding additional gas and steam turbines. Such plants boast efficiencies of up to 57%.

New generation combined cycle power plants will soon reach 60% efficiency. This is the efficiency expected by GE Power Systems from its H-System, the product of its project funded under the US DOE ATS programme. Such units are designed specifically for combined cycle operation, and the gas and steam turbines are closely coupled to ensure the maximum performance.

Micro turbines

A new trend within the gas turbine industry is the development of micro turbines, small gas turbines which can be used for power generation and cogeneration. These small turbines, with power generating capacities of between 10 and 100 kW can be installed in factories, office blocks or small housing developments.

Small turbines operate in exactly the same way as their larger relatives. However their small size makes them easy to integrate into a number of domestic or working environments. Emission performance is generally better than for larger gas turbines, and though efficiency is not comparable, when used for cogeneration of heat as well as electricity they can provide a competitive source of energy.

Environmental impact of gas turbines

One of the primary advantages of gas turbines is that they produce relatively little pollution, at least compared with coal-fired power plants. In the developed countries of the world where emission control has become a high-profile issue this has had a significant effect on the choice of technology for new generating capacity.

Most gas turbine power plants burn natural gas which is a clean fuel. Gas turbines are, anyway, extremely sensitive to low levels of impurities in the fuel, so fuel derived from other sources, such as gasification of coal or biomass, must be extensively cleaned before it can be burned in a gas turbine.

Even so, gas turbines are not entirely benign. They can produce significant quantities of NO_x, some carbon monoxide and small amounts of hydrocarbons. Of these, NO_x is generally considered the most serious problem.

Nitrogen oxides

NO_x emissions are generated during the combustion process. The amount of NO_x produced is directly related to the temperature at which combustion takes place. The higher the temperature, the more NO_x generated. And since gas turbine designers are pushing forever higher-turbine inlet temperatures in order to increase gas turbine efficiency, the problem of NO_x generation has become more acute with time.

It became apparent during the 1970s that development aimed at reducing the amount of NO_x generated in gas turbines would become necessary. One approach that met with some success was to inject water into the combustion chamber. This was eventually superseded by the use of dry low NO_x burners which control the mixing of fuel and air in such a way as to minimise the production of NO_x.

Early low NO_x burners did not prove as reliable as their manufacturers had hoped. Nevertheless the latter have pursued this line of development, with second generation low NO_x burners appearing at the beginning of the 1990s. The latest heavy gas turbine power plants can generally meet NO_x emissions targets in the range 15–25 ppm. New generation turbines, such as the H-Series from GE, expect to reach 9 ppm.

This level of NO_x and carbon monoxide emissions will meet the regulations in many parts of the world but not all. One of the countries that imposes more stringent limits is Japan. In order to meet these limits, a gas turbine has to be equipped with a selective catalytic reduction (SCR) system. This employs a metallic catalyst which stimulates a reaction between NO_x and added ammonia or urea, reducing the NO_x to nitrogen. SCR is expensive, but effective. A 2800-MW combined cycle power plant built by the Tokyo Electric Power Company at Yokohama in Japan employs SCR units to cut NO_x emission levels to less than 5 ppm.

Carbon dioxide

Gas turbines also produce large amounts of carbon dioxide. This is an unavoidable product of the combustion of natural gas. But a gas turbine power station produces proportionally less carbon dioxide than a conventional coal-fired power plant of similar capacity.

The reason for the better carbon dioxide performance is to be found in the composition of natural gas, which is primarily made up of methane. Each methane molecule contains one atom of carbon and four of hydrogen. When this burns in air it generates heat, one molecule of carbon dioxide and two molecules of water.

Coal is primarily composed of carbon. Therefore combustion of coal in air produces only carbon dioxide; it generates no water. The actual comparison is complicated by the amount of heat generated in each case and the efficiency of the two types of power station. But overall, the Electric Power Research Institute (EPRI) has estimated that a gas-fired power station produces around half the carbon dioxide of a coal-fired power station for each unit of electricity.

In the short term a switch from coal-fired to gas-fired power generation can, therefore, reduce carbon dioxide emissions significantly. Since carbon dioxide is a major contributor to the global greenhouse effect, switching is one strategy that is enabling some countries to meet (or attempt to meet) the emission targets of the Kyoto Protocol. In the long term, however, it seems probable that the continued use of natural gas as a power plant fuel will require some form of carbon dioxide capture. (Strategies to accomplish this have been outlined in Chapter 3.)

Carbon monoxide and particulates

Gas turbines can produce both carbon monoxide and small quantities of particulate material. Both result from incomplete combustion of natural gas. Levels of 10 ppm for both are typical.

Financial risks associated with gas-turbine-based power projects

The risks attached to electricity generating projects based on gas turbines fall into two main categories. There are those associated with gas turbine technology, and those associated with the cost and supply of fuel.

Technological risk

The gas turbine, as developed for aircraft propulsion, is an extremely reliable, efficient and robust machine. Safety and reliability are of prime importance to the airline industry and airline power units must meet exacting standards.

It can be assumed that aeroderivative gas turbines, based directly on aviation propulsion units, will show the same levels of reliability and efficiency as the machines from which they are derived provided no significant design modifications have been introduced. Design modification of these highly optimised machines is extremely costly, and design alteration for power generation applications makes little sense since its most likely service will be to increase costs. Consequently the risks associated with the use of aeroderivative gas turbines should be minimal. It is important, however, to clarify the relationship between the stationary machine and the aviation machine.

The same does not apply to heavy-frame gas turbines developed specifically for the power generation industry. These units do not have to meet the same exacting safety standards as the aviation units. Consequently they are generally not so thoroughly tested before entering service. Given the cost of a single 200–300 MW class gas turbine, it is perhaps not surprising to learn that some of the testing of these new heavy gas turbines has taken place in service. As a result there have been a number of instances of failure and the need for modification. Though the manufacturers are frequently coy about discussing such issues, it is clear that most if not all have been affected.

Part of this problem has arisen from the speed with which the market for heavy gas turbines for power generation has evolved and the high levels of competition this has engendered. Manufacturers may have learned from their recent experience, but even so a developer would be wise to establish the history of any gas turbine under consideration for a power project.

Fuel risk

Natural gas appears to be the fuel of the moment for the power generation industry. Demand is high, but supplies remain plentiful in most areas of the world. As a result, gas prices have remained low in most parts of the world (though there have been significant price fluctuations in the USA). This situation cannot be expected to continue.

The economics of gas-fired power generation rely heavily on low gas prices. Once gas prices start to rise coal-fired plants, even when fitted with costly emission-control systems, soon become more cost effective.

This presents a dilemma for companies planning to develop new gas-fired generating capacity. Over the short term it looks economically attractive – though recent experience in North America suggest that an open gas market can lead to rather large price fluctuations (see Table 4.2) – but longer-term uncertainties must remain. New technologies to gasify coal and use the gas generated to fire gas turbines offer one solution to this dilemma, but the technology has not been demonstrated widely enough to make coal gasification a realistic option in the near term. Besides gas turbines optimised for natural gas may not perform as well with coal gas.

The second factor is security of supply. In Europe and North America the construction of national and international gas transmission systems have made the supply and availability of gas stable. In most other parts of the world this gas infrastructure does not yet exist. Networks are being developed in Asia and South America but the cost of development is high, particularly as long distance gas transmission pipelines are often required. Under such circumstances, security is likely to be higher when the development of a gas-fired power project takes place close to a source of fuel.

A well-organised supply infrastructure will aid gas security but cannot ensure it. Western Europe is already being forced to import gas from remote regions of Russia and from Algeria to supplement its own dwindling resources. The USA is eying fields in Alaska to boost its resources. Such extended supply lines are vulnerable to both technical failure and terrorist attack, either of which could cripple gas supply in the future.

The cost of gas turbine power stations

In 1994 a report commissioned by the Center for Energy and Economic Development put the capital cost of a new combined cycle power plant to be built in the USA after the year 2000 at US$800/kW. In 2003 the US EIA estimated the overnight cost of a US combined cycle plant (in 2001$) which would start generating power in 2005 to be US$500–550/kW.[9] A simple cycle combustion turbine cost US$389/kW, The EIA estimated.

Comparing the 1994 figure with that for 2003 suggests that the cost of gas turbines has fallen during the intervening years. This is supported by anecdotal evidence. However US EIA figures from the end of the 1990s put the combined cycle cost at around US$440/kW, suggesting that if there was a fall in prices, that has now ended and prices are gradually rising.

It is difficult to obtain actual gas turbine costs because competition is fierce and manufacturers are loath to release prices. The only real source of data, therefore, is the published contract prices for actual projects. Table 4.3 collects together published data for a number of constructed or planned combined cycle power plants. While the published cost of a power plant can provide broad guidance only without much specific detail about each project and the elements included in the gross figure, they do indicate a lower limit of around $500/kW for the capital cost of a new combined cycle power station ordered in the late 1990s.

This estimate is supported by a Nortwest Power Planning Council report published in 2002 which estimated the overnight cost of a new combined cycle power plant to be around $565/kW, with an all-in cost of $621/kW. Depending on location, other estimates suggest that infrastructure

Table 4.3 *Combined cycle power plant costs*

	Capacity (MW)	*Cost (US$ million)*	*Cost/kW (US$)*	*Start-up*
UK (Teeside)	1875	1200	640	1993
Bangladesh (Sylhet)	90	100	1110	1995
India (Jegurupadu)	235	195	830	1996/1997
Malaysia (Lumet)	1300	1000	770	1996/1997
Indonesia (Muara Tawar)	1090	733	670	1997
UK (Sutton Bridge)	790	540	680	1999
Vietnam (Phu My 3)	715	360	500	2002
USA (Possum Point)	550	370	670	2003
Algeria	723	428	590	2006
Pakistan	775	543	700	–

Source: Modern Power Systems.

costs and land prices could as much as double this figure. Even so, the cost remains significantly lower than that of a coal-fired power plant.

In fact combined cycle power plants are the cheapest of all fossil-fuel-fired electricity generating stations to build. This makes them particularly attractive for countries with limited funds for power plant construction. They provide a cheap and fast addition to generating capacity, and the will be economical too, provided the cost charged for the power generated is sufficiently high to cover generating costs and loan repayments.

Operational and maintenance (O&M) costs for the gas turbine plant are competitive with coal. The EIA estimated that the variable O&M costs for a combined cycle plant (in 1996 prices) were 2.0 mills/kWh and the fixed O&M costs 15.0 mills/kWh. This compares with 3.25 and 22.5 mills/kWh for a conventional coal-fired power station.

Unlike a coal-fired power station, where much of the plant can be manufactured in the country where it is being built, a gas turbine is a highly technical and complex machine which can only be made by a limited number of manufacturers. This means that most countries of the world need to import all the gas turbines they use in electricity generating stations. Depending on the source of finance, this could make the gas-turbine-based power plant less attractive than the coal-fired alternative.

Such considerations have limited the use of gas turbines in developing countries that have not embraced private power production. But where this is permitted, the financing of the project becomes a matter for the project owner. Loans can often be raised in the country where a gas turbine is being manufactured, particularly from export agencies. National foreign reserves in the country where the plant is to operate are not required, for construction at least, making such a project more attractive.

With a gas turbine power station, capital cost represents but a small part of the total economic picture. More important is the cost of the fuel, which will be higher than the cost of fuel for the competitive coal-fired power station.

The total fuel bill over the lifetime of the power station has to be taken into account when determining whether a gas-fired project is more economical to build than one fired with an alternative fuel such as coal. The expected revenue is, of course, important too.

There are situations where power from a gas turbine plant can command a higher price than that from a coal-fired plant. Gas turbines can be started and stopped more easily, so they can be used to follow the demand curve, supplying peak power when demand is high. This is generally more highly valued than base-load power.

Thus the economics of the gas turbine plant are complex. Even so, many planners assume that is currently the cheapest cost option, quoting a generating cost of around $0.03/kWh. This figure depends on a number of assumptions, particularly discount rate over the lifetime of the plant. A recent challenge to conventional thinking put the generating cost in the range $0.05–$0.07/kWh.[10] That would make some renewable sources cheaper. Even so, there is no evidence yet for a waning in the popularity of the gas turbine for power generation.

End notes

1 World Energy Council, Survey of Energy Resources 2001.
2 International Energy Annual 2001, published by the EIA (March 2003).
3 Figures are quoted from European Natural Gas Supplies, Key Note, Power Economics (October 2002), pp. 24–27.
4 World Energy Council, Survey of Energy Resources, 2001.
5 Non-OEMs on their metal, James Varley, Modern Power Systems (May 2003), pp. 26–29.
6 CCGT plant progress: a Portuguese perspective, James Varley, Modern Power Systems (January 2003), pp. 25–26.
7 Humidified Gas Turbines, Arthur Cohn, presented at the Fourth Seminar on Combined Cycle Gas Turbines, British Institute of Mechanical Engineering, 1998.
8 Humidified Gas Turbines, Arthur Cohn, presented at the Fourth Seminar on Combined Cycle Gas Turbines, British Institute of Mechanical Engineering, 1998.
9 Annual Energy Outlook 2004, US Energy Information Administration.
10 Is gas really cheapest? Shimon Awerbuch, Modern Power Systems (June 2003) 17.

5 Combined heat and power

The production of electricity from fossil, biomass or nuclear fuels is an inefficient process. While some modern plants can achieve nearly 60% energy conversion efficiency, most operate closer to 30% and smaller or older units may reach only 20%. The USA, which has a typical mix of fossil-fuel-based combustion plants, achieves an average power plant efficiency of 33%. Other countries would probably struggle to reach even this level of efficiency.

Putting this another way, between 40% and 80% of all the energy burnt in power plants is wasted. The wasted energy emerges as heat which is dumped in one way or another. Sometimes it ends up in cooling water, but most often it is dissipated into the atmosphere. This heat can be considered as a form of pollution.

Some loss of energy is inevitable. Neither thermodynamic nor electrochemical energy conversion processes can operate even theoretically at 100% efficiency and practical conversion efficiencies are always below the theoretical limit. Hence, while technological advances may improve conversion efficiencies, a considerable amount of energy will always be wasted.

This energy cannot be utilised to generate electricity but it can still be employed. Low-grade heat can be used to produce hot water or for space heating[1] while higher-grade heat will generate steam which can be exploited by some industrial processes. In this way the waste heat from power generation can replace heat or steam produced from a high-grade energy source such as gas, oil or even electricity. This represents a significant improvement in overall energy efficiency.

Systems which utilise waste heat in this way are called *combined heat and power* (CHP) systems (the term co-generation is often used too). Such systems can operate with an energy efficiency of up to 90%. This represents a major saving in fuel cost and in environmental degradation. Yet while the benefits are widely recognised, the implementation of CHP remains low.

Part of the problem lies in the widespread preference for large central power stations to generate electricity. Such plants are sited to suit the demands of power network and of the power generation companies which own them. Rarely will there be a local use for the waste-heat energy such a plant produces. Only if generating capacity is broken down into smaller units, each located close to the source of demand, does it become possible to plan to use both electricity and heat.

History

Today breaking up generating capacity in this way is called *distributed generation*, but the concept is not new. Municipal power plants supplying district heating schemes are an early example of the same concept.

The potential for combining the generation of electricity and heat was recognised early in the development of the electricity generating industry. In the USA at the end of the nineteenth century, city authorities used heat from plants they had built to provide electricity for lighting to supply hot water and space heating for homes and offices. These district heating schemes were soon being replicated in other parts of the world.

In the UK engineers saw in this a vision of the future. Unfortunately uptake was slow and it was not until 1911 that a district heating scheme of any significance, in Manchester city centre, was developed.[2] Fuel shortages after World War I, followed by the depression made district heating more attractive in Europe during the 1920s and 1930s. Even so, take up was patchy.

By the early 1950s district heating systems had become established in some cities in the USA, in European countries such as Germany and Russia, and in Scandinavia. In other countries like the UK there was never any great enthusiasm for CHP and it gained few converts.

The centralisation of the electricity-supply industry must take some blame for this lack of implementation. Where a municipality owns its own power generating facility it can easily make a case on economic grounds for developing a district heating system. But when power generation is controlled by a centralised, often national body, the harnessing of small power plants to district heating networks can be seen as hampering the development of an efficient national electricity system based on large, central power stations.

Power industry structure is not the only factor. Culture and climate are also significant. So while the UK failed to invest in district heating, Finland invested heavily. Over 90% of the buildings in its major cities are linked to district heating systems and over 25% of the country's electricity is generated in district heating plants.

District heating was – and remains – a natural adjunct of municipal power plant development. But by the early 1950s the idea was gaining ground that a manufacturing plant, like a city, might take advantage of CHP too. If a factory uses large quantities of both electricity and heat, then installing its own power station allows it to control the cost of electricity and to use the waste heat produced, to considerable economic benefit. Paper mills and chemical factories are typical instances where the economics of such schemes are favourable.[3]

Technological advances during the 1980s and 1990s made it possible for smaller factories, offices and even housing developments to install CHP systems. In many cases this was aided by the deregulation of the power-supply industry and the introduction of legislation that allowed small producers to sell surplus power to the local grid. Since the middle of the 1990s the concept of distributed generation has become popular and this has also encouraged CHP.

Recent concern for the environment now plays its part too. Pushing energy efficiency from 30% to 70% or 80% more than halves the atmospheric emissions from a power station on a per kWh basis. Thus CHP is seen as a key emission-control strategy for the twenty-first century. But while environmentalists call for expanded use, actual growth remains slow. A government target in the UK of 10 GW of CHP power generating capacity by 2010, apparently achievable in 2000 looked set to fall short by close to 2.5 GW by the end of 2002.[4]

Applications

From single-home units to municipal power stations supplying heat and power to a city, from paper mills burning their waste to provide steam and heat to large chemical plants installing gas-turbine-based CHP facilities; CHP installations are as different as their applications are varied.

Ideally the heat and electricity from a CHP plant will be supplied to the same users. A homeowner might install a tiny fuel-cell-based power generation unit to replace the household boiler. The new unit will still supply household heating but will provide electricity too, with excess power perhaps being sold to the local grid.

On a larger scale a reciprocating engine burning natural gas could be used to supply both electricity and heat to an office building or a large block of apartments. And at the top end of the capacity scale, a municipal power plant based on a coal-fired boiler or a gas turbine can provide electricity for a city and heat for that city's district heating system. In all these cases the same users take both heat and power.

Similar opportunities exist in industry. Many processes require a source of heat and all industrial plants need electricity. Often the two can be combined to good economic effect, once the benefits are recognised. So where, in the past, a paper manufacturer would have installed a boiler to supply heat while buying power from the grid, now the same manufacturer is more likely to install a CHP plant.

Such instances represent the ideal but a good match of heat and power demand is not always possible. Sometimes electricity is required but no heat; and sometimes large quantities of heat or steam are needed but no electricity. With creative thinking CHP can be adapted to these situations too. For example a company that needs considerable quantities of steam but little electricity might built a gas turbine CHP plant designed to produce the quantity of heat it required, treating the electricity also produced as a by-product to be sold to the local grid.

Domestic heat consumption remains the challenge. Where district heating networks exist, a good balance between domestic heat and electricity demand is possible. But where these do not exist the only solution is either power stations meeting the electricity demand only, or domestic CHP systems. Finland offers one of the best examples with 25% of its electricity generated in CHP plants. In the USA in 2002 only 3.5% of power came from CHP plants.[5] Industrial CHP represented around 90% of the CHP capacity.

CHP technology

Most types of power generation technology are capable of being integrated into a CHP system. There are obvious exceptions such as hydropower, wind power and solar photovoltaic. But solar thermal power plants can produce excess heat and geothermal energy is exploited for CHP applications. Fuel cells are probably one of the best CHP sources while conventional technologies such as steam turbines, gas turbines and piston engine plants can all be easily adapted.

The type of heat required in a CHP application will often narrow the choice. If high-quality steam is demanded then a source of high-temperature waste heat will be needed. This can be taken from a steam-turbine-based power plant, it can be generated using the exhaust of a gas turbine and it can be found in a high-temperature fuel cell. Other generating systems such as piston engines are only capable of generating low-quality steam or hot water.

The way in which a CHP plant is to operate is another important consideration. Is it going to be required to provide base-load electricity generation or will it follow the load of the user who is installing it? If the plant will be required to load follow, then a power generating unit suited to that type of operation will be needed. The best for this purpose is either a fuel cell or a piston engine power plant. However if base-load electricity generation is intended, then a gas turbine or perhaps a steam turbine will offer the best solution.

These two can also provide steam supply flexibility. With a gas-turbine-based CHP plant, excess steam can be used to generate extra electricity. A steam-turbine-based system, meanwhile, will allow steam and electricity generation to be balanced to meet site demands.

The quantity of heat that will be available will also vary from technology to technology. Table 5.1 gives typical energy conversion efficiency ranges for modern fossil-fuel-burning power plants. Most of the energy not converted into electricity will be available as heat. Where more flexibility is required, it is possible to design a plant to produce less electricity and more heat that the efficiency figures in Table 5.1 suggest. Some technologies are amenable to this strategy. Others are not.

Table 5.1 *Power plant energy conversion efficiencies*

	Efficiency (%)
Conventional coal fired	38–47
Pressurised fluidised bed	45
Integrated-gasification combined cycle	45
Heavy gas turbine	30–39
Aeroderivative gas turbine	38–42
Gas turbine combined cycle	59
Fuel cell	36–60
Lean-burn gas engine	28–42
Slow-speed diesel	30–50

Most of the technologies employed in CHP plants have their own chapters in this book where detailed accounts of their operation can be found. In discussing these technologies here, consideration will only be given to factors of specific relevance to CHP. Please refer to other chapters for fuller accounts of each technology.

Piston engines

There are two primary types of piston engine for power generation, the diesel engine and the spark-ignition gas engine. Of these the diesel engine is the most efficient, reaching close to 50% energy conversion efficiency. The spark-ignition engine burning natural gas can achieve perhaps 42% efficiency but it is much cleaner than the diesel. Indeed it is impossible to obtain authorisation to use a diesel engine for continuous power generation service in some parts of the world.

There are four sources of heat in a piston engine: the engine exhaust, the engine jacket cooling system, the oil cooling system and the turbocharger cooling system (if fitted). Engine exhaust can provide low- to medium-pressure steam and the engine jacket cooling system can provide low-pressure steam. However the most normal CHP application would generate hot water rather than steam. If all four sources of heat are exploited, roughly 70–80% of the energy in the fuel can be utilised.

Piston engine power plants are available in sizes ranging from a few kW to 65 MW. These engines are particularly good at load following; a spark-ignition engine efficiency falls by around 10% at half-load while diesel engine efficiency barely drops over this range. There is no significant penalty in terms of engine wear for variable load operation. Piston engines can also be started quickly, with start-up times as short as 10 s typical.

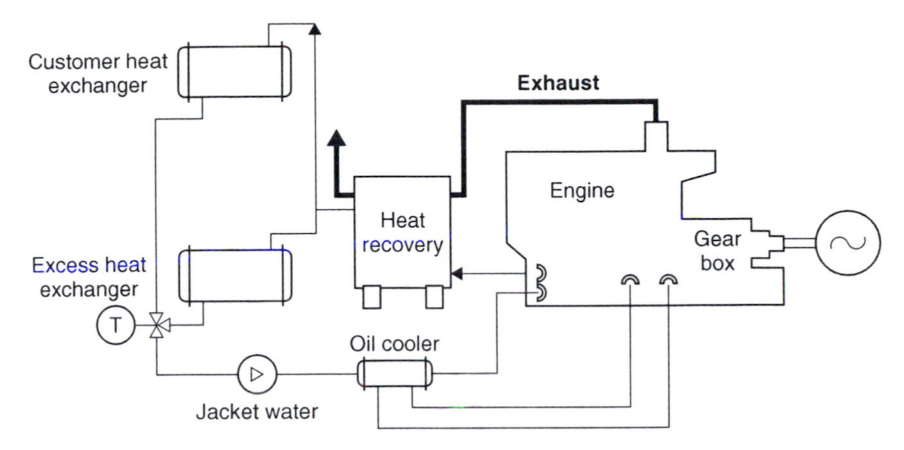

Figure 5.1 *Block diagram of piston-engine-based CHP system which is a closed-loop head-recovery system*

Applications for piston engine CHP plants include small offices and apartment blocks, hospitals, government installations, colleges and small district heating systems. Engines tend to be noisy, so some form of noise insulation is normally required. Emissions of gas engines can normally be controlled with catalytic-converter systems but diesel engines require more elaborate measures to control their higher nitrogen oxides and particulate emissions.

Steam turbines

A steam turbine is one of the most reliable units for power generation available. Modern large utility steam turbines have efficiencies of 46–47% but smaller units employed for CHP applications generally provide efficiencies of 30–42%. These turbines are usually simpler in design too. Steam turbines are available in virtually any size from less than 1 MW to 1300 MW.

A steam turbine cannot generate electricity without a source of steam. This is normally a boiler in which a fossil fuel or biomass fuel is burnt. However it can also be a waste-heat boiler exploiting the hot exhaust from a gas turbine. A steam turbine will normally be used in a CHP system only where there is a demand for high-quality, high-pressure steam for some industrial process.

There are a number of ways in which a boiler/steam turbine system can be used in such an application. One method is to take heat directly from the boiler to supply the process, with any surplus being used to drive the steam turbine. Alternatively steam can be taken from the boiler to the steam turbine and then from the turbine exhaust to the process. The pressure and temperature of the steam exiting the turbine can be tailored to suit the

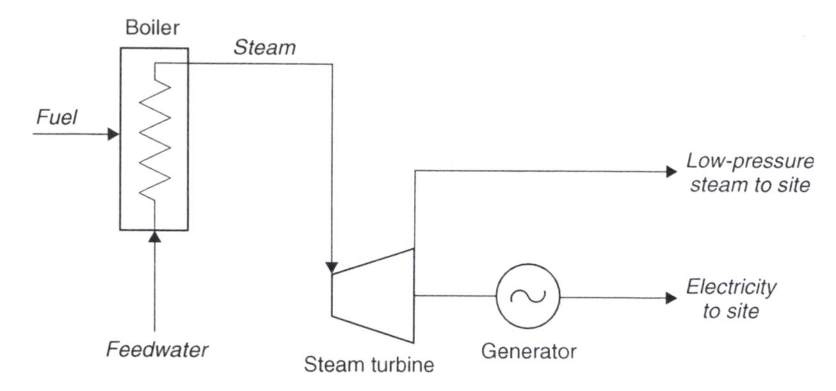

Figure 5.2 *Block diagram of steam turbine CHP system*

industrial demand. A third method is to extract steam from the turbine casing at a point before the exhaust. Combinations of all these methods are possible, so that the CHP system can be tuned for maximum efficiency.

The emissions from a steam turbine CHP system will be those of the boiler which generates the steam. Thus the emission-control measure will depend on whether the plant burns coal, wood or gas. Noise is unlikely to be a consideration since a steam turbine CHP system will only be used in an industrial environment.

Gas turbines

Unlike steam turbines, gas turbines burn fuel directly. Large industrial gas turbines operate with energy conversion efficiencies of up to 39% but smaller gas turbines, often derived from aeroengines, can operate at up to 42% efficiency. Gas turbine generating capacities range from 3 MW up to over 250 MW. Units of any size can be used in CHP systems and gas turbines are probably the cheapest prime movers available today. However they are best suited for continuous base-load operation. Regular output change can increase wear and maintenance significantly.

Most modern gas turbine installations burn natural gas, though some burn distillate. The heat output from a gas turbine is all found in its exhaust. This is a high-temperature source and it can be used to generate high-temperature, high-pressure steam. Hence a gas turbine will normally only be used in a CHP application where there is a need for high-quality steam. This steam will be generated in a waste-heat boiler attached directly to the turbine exhaust.

Two features provide the gas turbine with additional flexibility in CHP applications. Firstly the turbine is capable of generating steam of sufficient quality to power a steam turbine. This means that steam demand can be

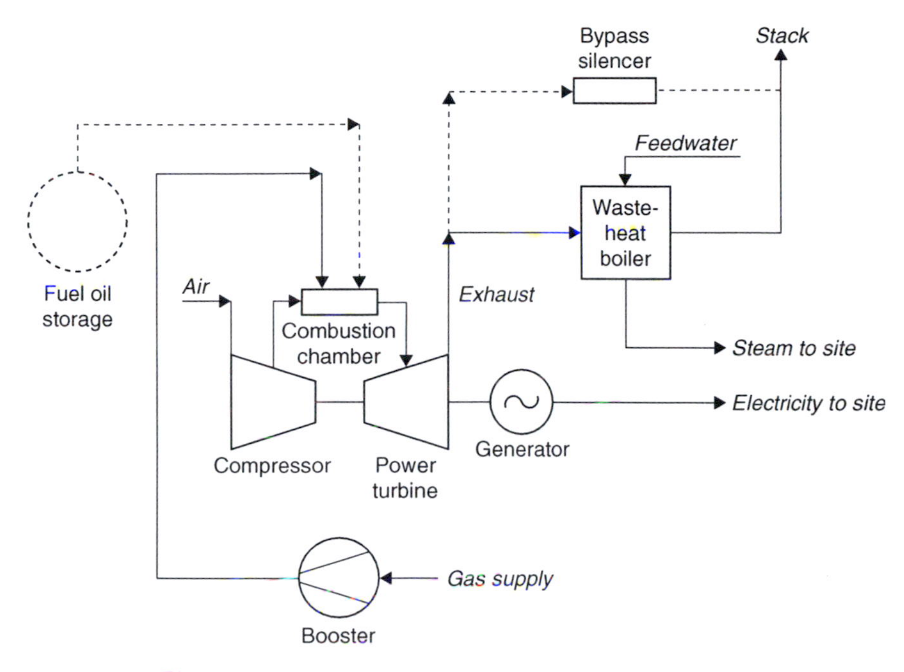

Figure 5.3 *Block diagram of gas-turbine-based CHP system*

balanced with electricity output by exploiting unwanted steam in a steam turbine to generate additional power. Secondly the exhaust from a gas turbine contains a considerable quantity of oxygen because the gas turbine combustion system employs an excess of air. This means that if necessary a waste-heat boiler can be fitted with a supplementary firing system to generate additional steam. This allows a gas turbine CHP system to be matched accurately to heat and electricity demand, allowing efficiencies of up to 90%.

Principal emissions from a gas turbine are nitrogen oxides. These can be controlled by using a special combustion system. Additional exhaust gas treatment may be necessary to meet more stringent environmental regulations. Like the steam turbine, a gas turbine CHP system is only likely to be considered in an industrial situation.

Micro turbines

Micro turbines are tiny gas turbines with capacities of from 25 to 250 kW. Many of these units are still in the development phase but some are now being deployed. These are often designed for CHP applications.

Micro turbines operate at extremely high speeds, often up to 120,000 rpm. Typical designs incorporate the turbine components and the generator on a single shaft. Bearings are air lubricated minimising wear. They can burn natural gas, gasoline, diesel and alcohol.

Micro turbine efficiency is low, in the 20–30% range. The exhaust heat can be used to generate low-pressure steam or hot water. The units are quiet relative to most engines so they can be run close to dwellings or in commercial environments. Their size range makes them suitable for commercial or light industrial applications. Multiple units operating in parallel can be used to increase capacity. Micro turbines are likely to become widely available by the end of the first decade of the twenty-first century.

Fuel cells

Fuel cells are electrochemical devices, like batteries, that convert a fuel directly into electricity. All fuel cells operate at an elevated temperature but some require very high temperatures while others work at only moderate temperatures.

Fuel cells are among the most efficient ways of converting fuel into electricity. Efficiencies range from 36% in operating units available today to a predicted 55–60% in high-temperature units under development. In CHP applications they can deliver up to 85% efficiency. Fuel cells are extremely good at load following, where their part-load loss of efficiency is minimal.

Low-temperature fuel cells such as the phosphoric acid fuel cell are well suited to distributed generation applications. The units are virtually noiseless so they can be positioned close to homes or offices without undue problem. Emissions are negligible too. However the cells are relatively expensive. These cells require hydrogen (normally derived from natural gas) and are easily poisoned, so the fuel must be very clean. Heat output is suitable for producing hot water but not steam.

Higher-temperature cells can burn natural gas directly without need for pre-treatment. These cells can produce both high-quality steam and hot water. Large units are likely to be deployed in a combined cycle configuration rather than for CHP but some companies are designing small high-temperature solid oxide fuel cells specifically for small commercial and even domestic environments.

While low-temperature fuel cells are commercially available, high-temperature fuel cells are still in their development phase, with the earliest models beginning to appear in commercial situations. Costs are high but these are expected to come down as economies of scale are realised. This technology is considered by many to be the best for future power generation, particularly in a hydrogen economy. It is particularly well suited to distributed generation CHP applications.

Nuclear power

A nuclear reactor is used in a power station as a source of heat energy, the heat being used to raise steam to drive a steam turbine. Thus in principle nuclear power can be used for CHP in exactly the same way as any other source of heat. While nuclear power is normally seen as best suited to base-load power generation in large-capacity plants, some attempts have been made to design and build nuclear plants to provide heat and power. In Russia and Eastern Europe some nuclear plants supply district heat and nuclear units have also been used to provide both electricity and heat for seawater desalination. The environmental concerns attached to nuclear generation have limited this type of use.

Environmental considerations

The primary environmental impact associated with a CHP plant will be a function of the technology employed in the plant. Atmospheric emissions will vary, depending on whether the plant employs a diesel engine, a gas turbine, a steam turbine with a biomass boiler or a nuclear reactor as the energy conversion system. Details of the impact of each of these technologies, and of all the others that are used in CHP installations, can be found in the chapters devoted to each elsewhere in this book. There are some environmental considerations that are of specific relevance to CHP and these will be considered here.

Noise

While noise generation may be a factor associated with all types of power generation, it is considered here because many CHP installations are designed for installation in commercial or urban domestic situations where any noise output is likely to be intrusive. Thus the noise output of a CHP plant will influence its usage.

The quietest of all CHP systems is probably the fuel cell. The actual electrochemical cells in a fuel cell operate silently. It has no generator, no turbine, no moving parts. However there is likely to be some noise associated with pumps and perhaps cooling systems. Designs intended for use close to homes or workplaces should be able to minimise noise to such an extent that it is no longer a consideration.

Micro turbines should also operate almost silently. These too are designed to be operated in close proximity to human activity.

Small piston-engine-based CHP systems are often intended for use in offices or for small district heating systems. However the engines are always noisy. They will normally require sound insulation and specially designed exhaust silencers for using in proximity with homes or offices. Underground or rooftop sites have often been employed to keep the units as isolated as possible.

Large piston engine plants, gas turbines and steam turbines are all relatively noisy and none is suitable for use close to housing or commercial units. These can all be used in large distributed generation applications but considerable attention to physical isolation of the site will be necessary.

Heat

The heat that is released into the environment by power stations can be classified as a form of pollution in the sense that it is a man-made waste which is being disposed of by this release. Generally its release is benign and this source of pollution is often ignored. However there are situations – such as using river water for power station cooling – where heat output can change the local environmental conditions significantly. This may be considered damaging.

A CHP plant reduces the immediate release of the waste heat from a power station and this must be considered an environmental benefit. Of course the heat will probably reach the environment eventually but when it does, it will do so in place of heat which would have been generated elsewhere. There is therefore a net reduction in the emission of heat.

Energy efficiency

The most significant impact of CHP is to increase, often dramatically, the energy efficiency of power generation. As we have seen, power plants at best can only convert 60% of the energy in the fuel they burn into electricity. Generally efficiency is much lower. On average probably 60–70% of all the fuel burned to generate electricity is dissipated as waste heat.

At the same time offices, homes, small commercial and large industrial plants are using electricity or fossil fuel to produce heat for space heating, for hot water and to provide the energy for chemical reactions.

When a CHP plant is installed, the heat it captures and utilises will replace one of these other sources of heat energy. So the fuel or electricity previously needed to produce this heat will be saved. This clearly represents a dramatic improvement in the use of energy and it is for this reason that CHP is considered by many to be a key element in future global energy strategies.

The reduced use of fuel as a consequence of CHP will also reduce atmospheric emissions. Here again the effect is dramatic. But how is one to quantify the savings made from the use of CHP?

Most commonly, manufacturers claim that a CHP plant is perhaps 80% or 90% efficient while the underlying power generating unit is only 30% efficient. Such figures are slightly misleading since they do not mean that 80% of the energy is being turned into electricity.

A more useful measure would be the amount of fuel saved by the use of CHP. For example if a power plant has a fuel to electricity conversion efficiency of 33%, and in a CHP installation 83% of the energy is captured, then 50% of the fuel energy is being captured as heat. Thus half the CHP plant fuel is replacing fuel otherwise needed to generate heat.

Finally one might also consider the economics of CHP. If a homeowner with a domestic boiler for central heating and hot water, and with a grid connection for electricity, installs a CHP unit which produces both, it will take a simple calculation to determine how much that homeowner has saved. Similar calculations apply to industrial installations. Normally such calculations show that the saving is significant.

Financial risks

The risk associated with investment in and construction of a CHP plant is primarily the risk associated with the underlying power generation technology. In most cases these technologies are proven and established and the technical risk should be extremely low. Newer technologies such as fuel cells are less well established and they do present a higher risk but for most applications the established technologies are going to be the most appropriate.

There can be some economic risk when CHP is being used in a situation where it has not previously been employed. Under these circumstances, heat and power demands need to be accurately assessed as well as the cost of installing and maintaining the CHP plant. A simple means of deriving the economic attractiveness of such an investment is to calculate the payback period. How long will it take to repay loans and recover the capital cost in savings from the new installation? If this is significantly less than the lifetime of the installation then the investment looks sound.

Cost of CHP

As with all power plant installations, the economics of CHP depend on the cost to install the power plant and the cost of the fuel. Table 5.2 lists the capital costs of the main CHP technologies. The figures in the table are applicable in the USA. Elsewhere, costs could vary. However they provide a useful means of comparing the different technologies.

As figures from Table 5.2 indicate, the gas turbine CHP plant is the cheapest overall, with installation costs in the range from $700 to $900/kW.

Table 5.2 *CHP costs*

	Capital cost ($/kW)	*O&M costs ($/kWh)*
Diesel engine	800–1500	0.005–0.008
Gas engine	800–1500	0.007–0.015
Steam turbine	800–1000	0.004
Gas turbine	700–900	0.002–0.008
Micro turbine	500–1300	0.002–0.010
Fuel cell	>3000	0.003–0.015

Source: California Energy Commission.[6]

Micro turbines may be able to undercut this in some instances but the different size ranges to which they apply makes it rare that they would be viable alternatives. Fuel cells are currently the most expensive of the six technologies listed. However their cost should come down by the end of the first decade of the twenty-first century.

Table 5.2 also includes typical operation and maintenance (O&M) costs. Again gas turbines and micro turbines offer the lowest-cost option but fuel cells, particularly high-temperature fuel cells should be extremely competitive when they finally reach the market.

Generation costs will depend on the fuel employed. The typical cost of electricity from a 100-kW CHP plant in the USA is $0.088/kWh while the cost of electricity from a 5-MW plant is $0.053/kWh.[7] In both cases these costs were competitive with commercial electricity rates in around 40% of the US market.

Viewing over a wider perspective, CHP as part of a distributed generation strategy offers additional economic gains. With distributed generation, transmission and distribution costs are lower, losses are reduced and the need for additional transmission and distribution capacity is reduced. Power system stability is improved and power dispatching becomes simpler.

Globally the gains are significant too. CHP means reduced fuel consumption which means reduced atmospheric emissions. While the costs of the latter as a result of health and environmental effects have been difficult to quantify, they are undoubtedly significant.

End notes

1 Heat can also be used to drive chillers and cooling systems. These are not considered separately here.
2 Combined Heat and Power in Britain, Stewart Russell in The Combined Generation of Heat and Power in Great Britain and the Netherlands: Histories of Success and Failure R1994: 29 (Stockholm: NUTEK, 1994).
3 Applications of this type are frequently designated co-generation rather than CHP. However the underlying premise is identical.
4 Review of CHP predictions to 2010, Ilex Energy Consulting, 2003.
5 US Energy Information Administration, Annual Energy Outlook, 2003.
6 Review of Combined Heat and Power Technologies, California Energy Commission, 1999.
7 Combined Cooling, Heat and Power Technology Overview, Keith Davidson, 2002 Energy Workshop and Exposition Hot Challenges, Cool Solutions Palm Springs, California (June 2002).

6 Piston-engine-based power plants

Piston engines or reciprocating engines (the two terms are often used interchangeably to describe these engines) are used throughout the world in applications ranging from lawn mowers to cars, trucks, locomotives, ships, and for power and combined heat and power generation. The number in use is enormous; the US alone produces 35 million each year. Engines vary in size from less than 1 kW to 65,000 kW. They can burn a wide range of fuels including natural gas, biogas, LPG, gasoline, diesel, biodiesel, heavy fuel oil and even coal.

The power generation applications of piston engines are enormously varied too. Small units can be used for standby power or for combined heat, and power in homes and offices. Larger standby units are often used in situations where a continuous supply of power is critical; in hospitals or to support highly sensitive computer installations such as air traffic control. Many commercial and industrial facilities use medium-sized piston-engine-based combined heat and power units for base-load power generation. Large engines, meanwhile can be used for base-load, grid-connected power generation while smaller units form one of the main sources of base-load power to isolated communities with no access to an electricity grid.

Piston engines used for power generation are almost exclusively derived from engines designed for motive applications. Smaller units are normally based on car or truck engines while the larger engines are based on locomotive or marine engines. Performance of these engines vary. Smaller engines are usually cheap because they are mass produced but they have relatively low efficiencies and short lives. Larger engines tend to be more expensive but they will operate for much longer. Large, megawatt scale engines are probably the most efficient prime movers available,[1] with simple cycle efficiencies approaching 50%.

There are two principle types of reciprocating engines, the spark-ignition engine and the compression or diesel engine. The latter was traditionally the most popular for power generation applications because of its higher efficiency. However it also produces high levels of atmospheric pollution, particularly nitrogen oxides. As a consequence spark-ignition engines burning gas have become the more popular units for power generation, at least within industrialised nations. A third type of piston engine, called the *Stirling engine,* is also being developed for some specialised power generation applications. This engine is novel because the heat energy used to drive it is applied outside the sealed piston chamber.

Piston engine technology

In its most basic form, the piston engine comprises a cylinder sealed at one end and open at the other end. A disc or piston which fits closely within the cylinder is used to seal the open end and this piston can move backwards and forwards within the cylinder. This it does in response to the expansion and contraction of the gas contained within the cylinder. The outside of the piston is attached via a hinged lever to a crankshaft. Movement of the piston in and out of the cylinder causes the crankshaft to rotate and this rotation is used to derive motive energy from the piston engine.

The manner in which the gas within the cylinder is caused to expand or contract defines the type of piston engine. In spark- or compression-ignition engines, valves are employed to admit a mixture of fuel and air into the sealed piston chamber where it is burnt to generate energy. Thus these engines are called *internal combustion engines*. In contrast the gas within a Stirling engine is caused to expand or contract by the application of heating and cooling from outside. This is called an *external combustion engine*.

Internal combustion engines form the major category of piston engines and these can be subdivided into spark- and compression-ignition engines. A further subdivision depends on whether the engine utilises a two- or a four-stroke cycle. The former is attractive in very small engines as it can provide relatively high power for low weight. For power generation, some very large engines also use a two-stroke cycle.[2] However most small- and medium-sized engines for power generation employ the four-stroke cycle.

The internal combustion variety of piston engine was developed in the latter half of the nineteenth century, although some primitive engines were in existence before that. Nikolaus Otto is generally credited with building the first four-stroke internal combustion engine in 1876. In doing so he established the principle still in use.

The Otto cycle engine employs a spark to ignite a mixture of air and, traditionally, gasoline[3] compressed by the piston within the engine cylinder. This causes an explosive release of heat energy which increases the gas pressure in the cylinder, forcing the piston outwards as the gas expands. This explosion is the source of power, its force on the piston turning the crankshaft to generate rotary motion.

The Otto cycle was modified by Rudolph Diesel in the 1890s. In his version, air is compressed in a cylinder by a piston to such a high pressure that its temperature rises above the ignition point of the fuel which is then introduced to the chamber and ignites spontaneously without the need for a spark.

The four-stroke cycle used by most of these engines derives its name from the four identifiable movements of the piston in the chamber – two of expansion and two of compression for each power cycle. With the piston at the top of its chamber, the first stroke in an intake stroke in which either air (diesel cycle) or a fuel and air mixture (Otto cycle) is drawn into the piston chamber (see Figure 6.1). The second stroke is the compression stroke during which the gases in the cylinder are compressed. In the case of the Otto cycle, a spark ignites the fuel–air mixture at the top of the piston movement creating an explosive expansion of the compressed mixture which forces the piston down again. This is the power cycle. In the diesel cycle fuel is introduced close to the top of the compression stroke, igniting spontaneously with the same effect. After the power stroke, the fourth stroke is the exhaust stroke during which the exhaust gases are forced out of the piston chamber. In either case a large flywheel attached to the crankshaft stores angular momentum generated by the power stroke and this provides sufficient momentum to carry the crankshaft and piston through the three other strokes required for each cycle.

As already noted, the piston is connected through a hinged lever to a crankshaft, this arrangement allowing rotary motion to be extracted from a linear movement. Normally four (or a multiple of four) pistons are attached to the crankshaft, with one of each set of four timed to produce a power stroke while the other three move through different stages of their cycles. The introduction of fuel and air, and the removal of exhaust is controlled by valves which are mechanically timed to coincide with the various stages of the cycle.

In a two-stroke engine, intake and exhaust strokes are not separate. Instead fuel is forced into the piston chamber (intake) towards the end of the power stroke, pushing out the exhaust gases through a valve at the top of the chamber. A compression stroke is then followed by ignition of the fuel and a repeat of the cycle.

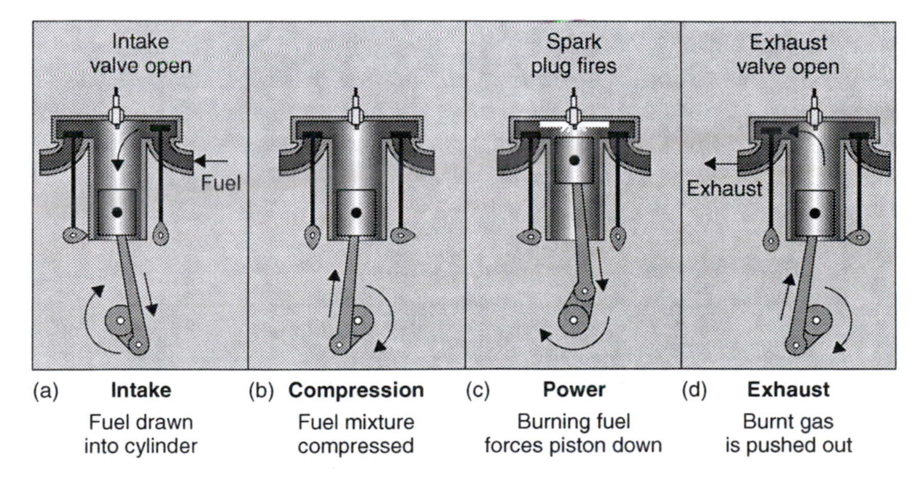

Figure 6.1 *The strokes of a four-stroke cycle*

Engine size and speed

The speed at which a piston engine operates will depend on its size. In general small units will operate at high speed and large units at low speed. However since in most situations a piston-engine-based power unit will have to be synchronised to an electricity grid operating at 50 or 60 Hz, the engine speed must be a function of one or other of these rates. Thus a 50 Hz high-speed engine will operate at 1000, 1500 or 3000 rpm while a 60 Hz machine will operate at 1200, 1800 or 3600 rpm.

Engines are usually divided into three categories, high-, medium- and slow-speed engines. High-speed engines are the smallest and operate up to 3600 rpm. The largest slow-speed engines may run as slow as 58 rpm. Typical speed and power ranges for each type of engine are shown in Table 6.1.

Engine performance varies with speed. High-speed engines provide the greatest power output as a function of cylinder size, and hence the greatest power density. However the larger, slower engines are more efficient and last longer. Thus the choice of engine will depend very much on the application for which it is intended. Large, slow- or medium-speed engines are generally more suited to base-load generation but it may be more cost effective to employ high-speed engines for back-up service where the engines will not be required to operate for many hours each year.

In addition to standby service or continuous output base-load operation, piston engine power plants are good at load following. Internal combustion engines operate well under part load conditions. For a gas-fired spark-ignition engine, output at 50% load is roughly 8–10% lower than at full load. The diesel engine performs even better, with output barely changing when load drops from 100% to 50%.

Spark-ignition engines

Spark-ignition engines can burn a variety of fuels including gasoline, propane and landfill gas. However the most common fuel for power generation applications is natural gas. Most are four-stroke engines and they are available in sizes up to around 6.5 MW.

Table 6.1 *Piston engine speed as a function of size*

	Engine size (MW)	*Engine speed (rpm)*
High speed	0.01–3.5	1000–3600
Medium speed	1.0–35.0	275–1000
Slow speed	2.0–65.0	58–275

Source: US Environmental Protection Agency.[10]

The spark-ignition engine uses a spark plug to ignite the fuel–air mixture which is admitted to each cylinder of the engine. In the simplest case this spark plug is located in the top of the cylinder and directly ignites the mixture within the cylinder. The fuel–air mixture within the cylinder will normally be close to the stoichiometric ratio required for complete combustion of the fuel although it may contain a slight excess of air (lean).

In larger, more sophisticated engines, the spark plug is contained within a pre-ignition chamber on top of the main cylinder. A fuel-rich mixture is ignited within the pre-ignition chamber and the flame shoots into the main chamber where it ignites the mixture there. The advantage of this system is that it allows the main mixture to contain a much larger excess of air over fuel. This results in a lower combustion temperature and this in turn reduces the quantity of nitrogen oxides produced.

The compression ratio of a spark-ignition engine (the amount by which the air–fuel mixture is compressed within the cylinder) must be limited to between 9:1 and 12:1 to prevent the mixture becoming too hot and spontaneously igniting, a process known as *knocking*. With natural gas, the engine efficiency varies between 28% (lower heating value, LHV)* for smaller engines and 42% (LHV) for larger engines. An engine tuned for maximum efficiency will produce roughly twice as much nitrogen oxides as an engine tuned for low emissions.

Many natural gas engines are derived from diesel engines. However because they must operate a modest compression ratios, they will only produce 60–80% of the output of the original diesel. This tends to make them more expensive than diesel engines. In practice this loss may be offset by longer life and lower-maintenance costs as a result of the derating of the engine and the cleaner fuel. Higher power can be achieved with a dual-fuel engine (see p.80).

Compression engines

Compression–ignition engines (diesel engines) use no spark plug. Instead they use a high-compression ratio to heat air within the cylinder to such a temperature that when fuel is finally admitted towards the end of the compression stroke, it ignites spontaneously. The compression ratio is normally in the range 12:1–17:1.

---

* The energy content of a fuel may be expressed as either the higher heating value (HHV) or the lower heating value (LHV). The higher heating value represents the energy released when the fuel is burned and all the products of the combustion process are then cooled to 25°C. This energy then includes the latent heat of vapourisation released when any water produced by combustion of, for example, natural gas, is condensed to room temperature. The lower heating value does not include this latent heat and is hence around 10% lower than the higher heating value in the case of natural gas.

The efficiency of the diesel engine ranges from 30% (higher heating value, HHV) for small engines to 48% (HHV) for the largest engines. Research should push this to 52% (HHV)[4] within the next few years. Diesel engines can be built to larger sizes than spark-ignition engines, with high-speed diesels available in sizes up to 4 MW and slow-speed diesels up to 65 MW. Large slow-speed engines can have enormous cylinders. For example, a nine-cylinder, 24 MW engine used in a power station in Macau has cylinders with a diameter of 800 mm.

The combustion temperature inside a compression-ignition engine cylinder is much higher than within a spark-ignition engine cylinder. As a consequence, nitrogen oxide emissions can be 5–20 times greater than from an engine burning natural gas. This can prove a problem and emission reduction measures may be required to comply with atmospheric emission regulations.

Diesel engines can burn a range of diesel fuels including both oil-derived fuels and biofuels. Smaller, high-speed engines normally use high-quality distillate but the large slow-speed engines can burn very low-quality heavy fuel oils which require a much longer combustion time to burn completely. These fuels tend to be dirty and plants burning them usually require additional emission mitigation measures.

Dual fuel engines

A dual fuel engine is an engine designed to burn predominantly natural gas but with a small percentage of diesel as a pilot fuel to start ignition. The engines operate on a cross between the diesel and the Otto cycles. In operation, a natural gas–air mixture is admitted to the cylinder during the intake stroke, then compressed during the compression stroke. At the top of the compression stroke the pilot diesel fuel is admitted and ignites spontaneously, igniting the gas–air mixture to create the power expansion. Care has to be taken to avoid spontaneous ignition of the natural gas–air mixture, but with careful design the engine can operate at close to the conditions of a diesel engine, with a high-power output and high efficiency, yet with the emissions close to those of a gas-fired spark-ignition engine.

Typical dual fuel engines operate with between 1% and 15% diesel fuel. Since a dual fuel engine must be equipped with diesel injectors, exactly as if it were a diesel engine, a dual fuel engine can also burn 100% diesel if necessary, although with the penalty of much higher emissions.

Stirling engines

Whereas fuel combustion takes place within the cylinders of an internal combustion engine, the heat energy used to drive a Stirling engine is

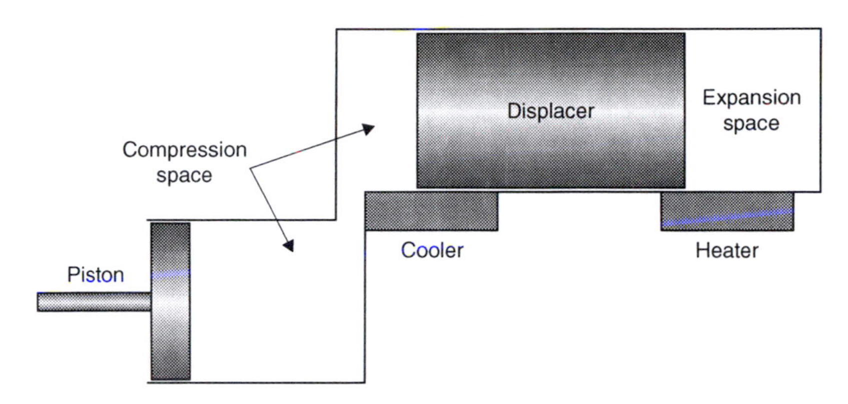

Figure 6.2 *The Stirling engine*

applied outside the cylinders which are completely sealed. The engine was designed by a Scottish Presbyterian minister, Robert Stirling, who received his first patent in 1816. The original engines used air within the cylinders and were called *air engines* but modern Stirling engines often use helium or hydrogen.

A normal Stirling engine has two cylinders, an expansion cylinder and a compression cylinder. The two are linked and heat is applied to the expansion cylinder while the compression cylinder is cooled. Careful balancing of the system allows the heat energy to be converted into rotational motion as in an internal combustion engine.

The great advantage of the Stirling engine is that the heat energy is applied externally. Thus the energy can, in theory, be derived from any heat source. Stirling engines have been used to exploit solar energy and for biomass applications. However their use is not widespread. Typical engines' sizes in use and development range from 1 to 150 kW.

Co-generation

when a heat engine is used to generate electricity, a large part of the energy supplied to the engine in the form of fuel emerges as waste heat. This applies equally to gas-turbine-, steam-turbine- and piston-engine-based power plants. If this heat can be captured it can be utilised for space heating, water heating or for generating steam, thus making much more efficient use of the fuel.

The efficiency of piston-engine-based power generation varies from 25% for small engines to close to 50% for the very largest engines. Thus between 50% and 75% of the fuel energy emerges as waste heat. In the case of an internal combustion engine, there are four primary sources of waste heat, the engine exhaust, engine case cooling water, lubrication oil cooling water and, where one is fitted, turbocharger cooling.[5] Each of these can be used as a source of heat in a combined heat and power system. And since the cooling systems are needed to remove heat from the engine and prevent it overheating, internal combustion engines can be quite simply converted into combined heat and power systems.

The exhaust gas contains between 30% and 50% of the waste heat from the engine. This can be used to generate medium-pressure steam if required. Otherwise it can be used to generate hot water. The main engine case cooling system can capture up to 30% of the total energy input. This will normally be passed through a heat exchanger to provide a source of hot water although in some cases it can be used to produce low-pressure steam as well. Engine oil and turbocharger cooling systems will provide additional energy that can be used to supply hot water.

If all the heat from the exhaust and the cooling systems of an engine is exploited, around 70–80% of the fuel energy can be used. However this falls dramatically where there is no use for hot water. Engine exhaust gases can also be used directly for drying in some situations.

An internal combustion engine must be fitted with cooling systems whether the waste heat is exploited or not, so the use of these systems in combined heat and power applications offers a logical extension of their application. Systems based on small engines can provide power, space heat and hot water to homes and commercial offices while large engines can produce power and process heat for small industrial operations. The economics of these systems can be quite favourable where there is a use for the waste heat. As a consequence they have become extremely popular. In the USA in 2000, for example, there were 1055 engine-based combined heat and power (CHP) systems in operation with an aggregate generating capacity of 800 MW.[6]

Combined cycle

The waste heat from the exhaust of an internal combustion engine is generally hot enough to generate medium-pressure steam. In the case of small engine installations, steam production is not normally an economical option unless there is a local use for the steam. In the case of a large diesel installation, however, the engine exhaust can be used to generate steam in a boiler, steam which can drive a steam turbine to produce additional energy. This forms the core of a diesel-engine-based combined cycle plant.

Typical of this sort of application is a generating plant which was installed in Macau in 1987.[7] This plant was equipped with a slow-speed diesel engine with a capacity of 24.4 MW. The engine exhaust was fitted with a waste-heat boiler and steam turbine which could generate an additional 1.34 MW when the engine was operating at full power, thus contributing around 5% of the plant output. As a result of this and other measures a fuel-to-electricity conversion efficiency of close to 50% was achieved.

Large engines of this type are frequently derived from marine engines and the original engines upon which they are based are not normally optimised for combined cycle operation. In particular, the cooling system is designed to keep the engine as cool as possible. For best combined cycle performance, however, it is preferable to run the engine as hot as possible.

Combined cycle performance of a large diesel engine can be improved by modifying engine components such that they can operate continuously at a higher temperature. This results in a higher-temperature exhaust which can be used to generate higher-quality steam to drive a steam turbine. With these measures it may be possible to achieve a fuel to electricity conversion efficiency of close to 55%. However the additional expense of the waste-heat recovery and steam turbine will only prove cost effective if the engine is to be used for base-load operation.

Environmental considerations

Piston engine power units generally burn fossil fuels and the environmental considerations that need to be taken into account are exactly the same considerations that affect all coal-, oil- and gas-fired power plants; the emissions resulting from fuel combustion. In the case of internal combustion engines the main emissions are nitrogen oxides, carbon monoxide and volatile organic compounds (VOCs). Diesel engines, particularly those burning heavy diesel fuel will also produce particulate matter and some sulphur dioxide.

Nitrogen oxides are formed primarily during combustion by a reaction between nitrogen and oxygen in the air mixed with the fuel. This reaction takes place more rapidly at higher temperatures. In lean-burn gas engines where the fuel is burned with an excess of air, temperatures can be kept low enough to maintain low nitrogen oxide emissions. The diesel cycle depends on relatively high temperatures and as a consequence of this produces relatively high levels of nitrogen oxides. Table 6.2 compares emissions from the two types of engines.

Table 6.2 *Emissions of nitrogen oxides from internal combustion engines*

	Emissions (ppmv)	*Emissions (g/kWh)*
High- and medium-speed diesel	450–1800	7–20
Natural gas burning spark-ignition engine	45–150	1–3

Source: US Environmental Protection Agency.

When the fuel in an internal combustion engine is not completely burned the exhaust will contain both carbon monoxide and some unburnt hydrocarbons. Carbon monoxide is poisonous and its levels should be minimised. The unburnt hydrocarbons are classified as VOCs, and both their emissions and those of carbon monoxide are controlled by legislation.

Natural gas contains negligible quantities of sulphur so gas engines produce no sulphur dioxide. Diesel fuels can contain sulphur. Small- and medium-sized diesel engines generally burn lighter diesel fuels which contain little sulphur. Larger engines can burn heavy residual oils which are comparatively cheap but which often contain significant levels of sulphur. As sulphur can damage the engine, it is normal to treat this type of fuel first to remove most of the sulphur.

Liquid fuels may produce particulate matter in an engine exhaust, the particles derived from ash and metallic additives. Incomplete combustion of heavy fuel can also lead to the emission of particulate matter.

Emission control

The most serious exhaust emissions from a piston engine are nitrogen oxides. Engine modifications that reduce the combustion temperature of the fuel, such as the use of a pre-combustion chamber and lean fuel mixture described above, offer the best means of reducing these emissions. Natural gas engines designed to burn a very lean fuel (excess air) provide the best performance. Diesel engines present a greater problem but water injection can reduce emission levels by 30–60%.

Where these measures are insufficient to keep emissions below regulation levels, exhaust gas treatment will be necessary. For small gasoline engines a simple catalytic converter of the type used in automobiles is often the most effective solution. This type of system cannot be used with diesel or with lean-burn engines although new catalysts for use with lean-burn engines are currently under development. Where it can be used, the catalytic converter will reduce nitrogen oxide emissions by 90% or more.

The alternative is a selective catalytic reduction system. This also employs a catalyst, but in conjunction with a chemical reagent, normally ammonia or urea, which is added to the exhaust gas stream before the emission-control system. The reagent and the nitrogen oxides react on the catalyst, and the nitrogen oxides are reduced to nitrogen. This type of system will reduce emissions by 80–90%. However care has to be taken to balance the quantity of reagent added so that none emerges from the final exhaust to create a secondary emission problem.

The emission of carbon monoxide, VOCs and some particulate matter can be controlled by ensuring that the fuel is completely burnt within the engine. Careful control of engine conditions and electronic monitoring systems can help maintain engine conditions at their optimum level. Exhaust gas catalytic oxidation systems can also be used to keep levels below prescribed limits. Old engines as they become worn can burn lubrication oil, causing further particulate emissions.

Sulphur emissions are only likely to be met with large diesel engine power plants burning heavy fuel oil. Some of these oils can contain as much as 3.5% sulphur. Normally this sulphur can be removed by pretreating the fuel. However in the worst case, a sulphur capture system can be fitted to the exhaust system. This adds to both capital and maintenance costs, and affects plant economics.

Carbon dioxide

The combustion of all carbon-base fuels results in carbon dioxide. This is as true of natural gas, oil or biodiesel as it is of coal. Coal is predominantly composed of carbon and it produces the greatest amount of carbon dioxide for each unit of heat energy. Liquid and gaseous fuels normally produce less. However in all cases significant emissions are inevitable.

Emission levels can be minimised by operating the engine at the highest efficiency possible. Use of waste heat increases efficiency and so helps minimise emission. Bio-derived fuels are generally considered carbon dioxide neutral since although their combustion generates carbon dioxide, production of more fuel results in the capture of the carbon dioxide again.

The only method of physically reducing carbon dioxide emission from fossil fuel combustion is to capture it and store it. Technology to achieve this is being developed but the cost is likely to be extremely high. It seems unlikely that it will ever be an economic option for small piston engines.

Financial risks

The technology used in the construction of all types of piston engines is mature and the nature of the processes involved are well understood. Improvements are continually made but these are minor in nature. Overall the performance and reliability of a piston engine should fall within well-established boundaries.

Performance, both in terms of overall efficiency of operation, reliability and lifetime, should be guaranteed by the manufacturer of a unit. Continuous operation of an engine represents the least onerous regime and performance under these circumstances should be predictable; continual starting and stopping, as encountered in transport applications, puts much greater strain on the machine.

By far the greatest risk attached to the operation of a piston engine power plant is related to fuel supply. Oil prices can be particularly volatile, but gas prices are likely to become subject to the same movements in price in the future. The development of a large piston engine power plant will usually include a long-term fuel supply agreement. However the operation of many smaller units will depend on the purchase of fuel at the current market price. This should always be taken into account when planning a project.

There is evidence that oil and gas supplies will face increasing pressure over the next few decades. This is likely to have an adverse effect on piston engine power plants. If such plants are to continue to serve as power generation units they will eventually need an alternative source of fuel such as hydrogen or biofuel.

Costs

The capital cost of a piston engine power plant generally depends on unit size. Small engines are generally mass produced and cheaper than their larger relatives. However this is often offset by higher installation costs. Thus typical total plant costs for a 100 kW generator unit is $1515/kW while a 5000 kW installation costs $919/kW.[8]

While plants in the 100–5000 kW capacity range are based on standard components, large piston engine power plants generally have a cost structure more like that of a gas turbine power plant. Table 6.3 lists the costs of a number of large diesel-engine-based power stations. These plants were built in different countries, using different engine configurations, and yet the unit cost of the plants all fall within a remarkably narrow range of $1100–1300/kW.

Maintenance costs vary with engine size and type. Small, high-speed engines generally require the most frequent maintenance while larger engines can run for much longer periods without attention. Engine oil

Table 6.3 *Typical large diesel power plant costs*

Project	*Capacity (MW)*	*Cost ($million)*	*Cost/kW ($)*	*Start-up*
Kohinoor Energy, Pakistan	120	140	1167	1997
Gul Ahmed Energy Co, Pakistan	125	138	1104	1997
Jamaica Energy Partners	76	96	1263	–
APPL, Sri Lanka	51	63	1235	1998
IP, Tanzania	100	114	1140	1998
Kipevu 2, Kenya	74	84	1135	2002

Source: Modern Power Systems.

monitoring systems are often used, particularly in large engines, to monitor wear rates. The US EPA found that maintenance cost varied between \$0.007/kWh and \$0.02/kWh for engines in the 100 kW–5000 kW range, with the smallest engines incurring the highest costs.[9]

Capital cost is a significant factor in the cost of electricity from a piston-engine power plant but the fuel cost is normally more important. On a cost per kWh basis, gas engines up to 5 MW will normally be able to compete with gas turbine units of similar size, the higher efficiency of the reciprocating engine in simple cycle mode providing a slight edge in many cases. Such engines are becoming increasingly popular for distributed generation applications. The advantage of reciprocating engines may extend to engines of up to 50 MW in capacity under certain conditions. For example, where the power plant is required to load follow, or at high altitude, the reciprocating engine has a significant advantage.

Diesel engines have a long history of use in supplying power to remote communities or isolated commercial facilities. Generation costs under these circumstances can be high as the fuel has to be shipped to the site, adding transport costs. Often, although, the diesel unit is the only viable source of power. Renewable energy systems such as wind, solar and small hydropower now offer an alternative to diesel in some cases.

Large, slow-speed diesel engines burning poor quality residual oil are generally a cost-effective source of electricity provided the fuel is available. Their use is, however, restricted by fuel supply.

The other major application of piston engines is for combined heat and power. Where there is a use for the heat supplied by a unit, this is normally an extremely cost-effective option. Such considerations are also encouraging the installation of distributed generation units in increasing numbers in developed regions of the world such as North America and Europe. As a consequence of this and other incentives, US orders for stationary engines grew by 68% in the year to June 2001, with natural gas-fired engine orders up by 95%. This is a trend which is expected to continue in the near future.

End notes

1 Slow-speed engines are the most efficient engines for converting fuel energy via heat into rotary motion to generate electricity. Fuel cells, which turn chemical energy directly into electrical energy, can be more efficient.
2 The two stroke is efficient and extremely tolerant of poor quality fuels.
3 Otto's engine probably burnt powdered coal but gasoline soon became the preferred fuel.
4 Technology Characterization: Reciprocating Engines, US Environmental Protection Agency, 2002.
5 Refer *supra* note 4.

6 A turbocharger is sometimes used to compress air before it is admitted into the cylinder of an internal combustion engine. This can lead to improved performance by generating greater power from the engine.
7 PA Consulting Independent Power Database, Energy Nexus Group. Figures are quoted in Technology Characterization: Reciprocating Engines, US Environmental Protection Agency, 2002.
8 The plant was built by Burmeister and Wain Scandinavian Contractor.
9 Refer *supra* note 4.
10 Refer *supra* note 4.

7 Fuel cells

The fuel cell is an electrochemical device, closely related to the battery, which harnesses a chemical reaction between two reagents to produce electricity. A battery is usually intended as a portable or self-contained source of electricity and it must carry the reagents it needs with it. Once they are exhausted the battery can no longer supply electricity. A fuel cell, by contrast, is supplied with reagents externally. So long as these reagents are made available, the cell will continue to provide power.

In addition to this difference between the fuel cell and a battery, there is something special about the chemical process the fuel cell harnesses to generate electricity. It consumes hydrogen and oxygen (usually supplied as air) and the only product of the reaction is water. The simplicity of this energy producing reaction and its inherent cleanliness makes the fuel cell an extremely attractive proposition from an environmental perspective.

Of course there is no large-scale source of hydrogen today, so fuel cells have to make do with hydrogen generated from natural gas in a chemical-reforming process. For now, this somewhat tarnishes the environmental credentials of the system. Nevertheless it can still provide an environmentally attractive source of electricity.

There are already a multitude of ways of generating electricity from a fossil fuel such as natural gas; why develop another? The answer is that there is a major difference between a fuel cell and these other electricity generating plants. Fossil fuel power stations which employ gas turbines, steam turbines or piston engines are all reliant on the thermodynamics of a heat engine. This limits the maximum theoretical efficiency that such devices can achieve.[1] The fuel cell, by contrast, is limited by electrochemical conversion efficiency. Thus, while the highest efficiency a modern simple cycle heat engine can achieve is around 50%, the best fuel cell can convert 70% of the fuel energy into electricity.[2]

The fuel cell has other advantages too. The cell itself has no moving parts and can operate for long periods without maintenance, far longer than any turbine- or engine-based generating system. The absence of moving parts makes them inherently quiet too (although this is limited by the use of mechanical pumps which do generate noise) and they emit relatively low levels of pollution compared to other types of generating system based on fossil fuel.

With so much going for them, why are there no fleets of fuel cell power plants today? The answer is cost. While the fuel cell principle has been known since the first half of the nineteenth century, development of a cheap version of the device has proved extremely challenging. As a result the first commercial fuel cells only appeared in the early 1990s and these were never competitive. Much research and investment has taken place since then and new generations of fuel cells are expected in the next 4 or 5 years which will be much more competitive.

The fuel cell principle

If an voltage is applied to water by placing two electrodes into the liquid and attaching a battery to them, the voltage induces a chemical reaction; hydrogen is produced at one electrode and oxygen at the other.

In 1839, Sir William Grove observed that this process, known as *hydrolysis*, will also go backwards. If two specially selected electrodes are placed in water containing an acid and gaseous hydrogen and oxygen provided, one to each, hydrogen will react at one electrode, and oxygen at the other, producing an electrical voltage between the electrodes. This is the basis of operation of the fuel cell.

Although the principle was known in 1839, it was not until a century later that Francis Bacon began to develop practical fuel cells. In the late 1950s, Pratt and Whitney Aircraft Corporation licensed Bacon's technology and developed it for use in the US space programme. As a result, the Gemini and Apollo space programmes and the space shuttle have all used fuel cells to generate electricity.

Work on a variety of fuel cells continued through the 1960s, 1970s and 1980s, culminating in the first commercial cell in 1992. Since then work has accelerated and several different types of cell will become commercially available during this first decade of the twenty-first century.

Fuel cell chemistry

The fuel cell belongs to a branch of chemistry called *electrochemistry*. This explains how electricity can be derived from a chemical reaction.

In nature certain materials will react with one another spontaneously. Sulphuric acid will dissolve metals. The two components of an epoxy glue such as Araldite react when mixed to form a tough adhesive. Natural gas burns in air to produce heat.

The particular reaction that is important to the fuel cell is that between hydrogen and oxygen. As the Hindenburg airship disaster graphically illustrated to the world in 1937, hydrogen will burn in oxygen or air releasing a vast quantity of energy in the form of heat. In the fuel cell this energy does not appear as heat. Instead it emerges as electrical energy.

The reaction between hydrogen and oxygen can be expressed by a simple chemical formula:

$$2H_2 + O_2 = 2H_2O$$

Two hydrogen molecules and one oxygen molecule react to create two molecules of water. In fact the reaction can be broken down into two separate halves, one involving hydrogen and one involving oxygen. The hydrogen half involves the hydrogen molecule, H_2, splitting into two hydrogen atoms, H, and each of these releasing an electron to form a positively charged hydrogen ion, a proton:

$$H_2 = 2H^+ + 2e^-$$

On the oxygen side, the oxygen molecule, O_2, also splits into two oxygen atoms, O, and each of these absorbs two electrons released from two hydrogen atoms to produce a doubly negatively charged oxygen ion, O^{2-}:

$$O_2 + 4e^- = 2O^{2-}$$

Now the negatively charged oxygen ion will attract two positively charged hydrogen atoms and they will coalesce to form a water molecule, H_2O:

$$O^{2-} + 2H^+ = H_2O$$

Then the reaction is complete.

When hydrogen burns in air, the various steps of the reaction occur in the same place at the same time. But in a fuel cell the hydrogen and oxygen are not allowed to mix.

In the fuel cell the hydrogen is supplied to one electrode of the cell and oxygen to the other. The two electrodes are separated by a material called the *electrolyte*. This electrolyte is impermeable to the gases. It will not conduct electricity either. What it will do is conduct positively charged hydrogen ions.

So at the hydrogen electrode (called the *anode*) the hydrogen molecules first separate into atoms and then release electrons to form positively charged ions. Only then can the hydrogen cross the electrolyte boundary and reach the oxygen at the second electrode. But at that electrode (called the *cathode*) it will find only oxygen atoms and molecules; these still need to pick up the electrons that the hydrogen atoms released at the anode of the cell if they are to complete the reaction.

Now electrons are what produce an electrical current. If a wire is connected between the anode and the cathode of the cell, the electrons will rush from one to the other in order to complete the reaction. Put a small electric light bulb in the circuit and it will glow, proving that there is indeed a current flowing. The fuel cell allows the hydrogen to get to the oxygen one way, but forces electrons to take a different route. That is how it works.

The electrolyte is the key to the operation of a fuel cell and the different types of fuel cell under development are normally identified by the electrolyte each uses. In the illustration below the electrolyte only allowed

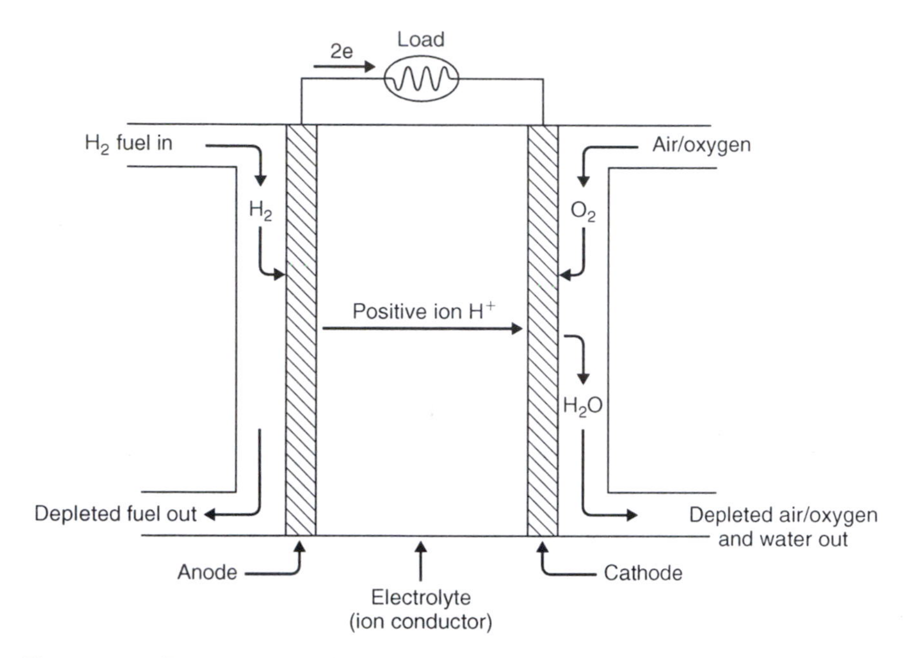

Figure 7.1 *The principle of the fuel cell*

charged hydrogen atoms to pass. This is the basis of several fuel cells. Others will only allow charged oxygen atoms to pass and in yet other cases the electrolyte is permeable only to a more complex charged molecule. And in every case the electrolyte must not allow electrons to pass through it from one electrode to the other. That would create a short circuit.

Catalysts

The description of the operation of a fuel cell above is a simplification because it omits one key feature of the reaction between hydrogen and oxygen. Although hydrogen atoms and oxygen atoms will react spontaneously to form water, both hydrogen and oxygen are found (at room temperature) in the molecular forms H_2 and O_2. These will not react spontaneously and the hydrogen and oxygen molecules must be split before the reaction will proceed.

One method of splitting the molecules is to raise their temperature. Thus a flame will split sufficient of the molecules to start the reaction which then generates enough heat spontaneously to keep the reaction going. Some fuel cell designs use high temperatures too.

The alternative is to use a catalyst. A metal such as platinum will promote the splitting of both hydrogen and oxygen molecules at low temperatures and the resulting atoms will then react in a fuel cell. However platinum is very expensive. This has a significant effect on the cost of low-temperature fuel cells.

Hydrocarbon gas reformation

The simplest fuel cell 'burns' hydrogen and oxygen in order to generate power. But hydrogen is not a readily available fuel. Fortunately hydrocarbon gases such as natural gas or gas generated from biomass (this contains a large quantity of methane) can easily be converted into a mixture of hydrogen and carbon dioxide. This will also form a suitable fuel for a fuel cell.

The conversion is usually carried out as a two-stage process. In the first stage methane is mixed with water vapour and passed over a catalyst at high temperature where it reacts to produce a mixture of hydrogen and carbon monoxide. A second reaction, called a *water shift reaction*, is then carried out during which additional water vapour reacts with the carbon monoxide to produce more hydrogen and carbon dioxide.

The second stage is extremely important for fuel cells because the catalysts in low-temperature cells are sensitive to carbon monoxide poisoning. In consequence, virtually all the carbon monoxide must be removed from the fuel before it is fed into the fuel cell. Fuel cells are sensitive to any sulphur impurities too and these must be scrupulously removed.

While natural gas is the most convenient source of hydrogen for a fuel cell today, other fuels can also be exploited. Methanol can also be converted into a hydrogen-rich gas using a reforming process, as can gasoline, though the latter requires an extremely high temperature (800°C). Both these processes are of interest to the automotive industry.

Since reforming of all these fuels takes place at a relatively high temperature, low-temperature fuel cells usually need an external reformer to supply their fuel. The conditions inside a high-temperature fuel cell are sufficient for the reforming to take place within the cell, simplifying system design.

It is important to note that while a fuel cell burning hydrogen and oxygen produces no carbon dioxide, most fuel cells will generate carbon dioxide because they derive their hydrogen from a fossil fuel. When methane is converted into hydrogen it generates exactly the same amount of carbon monoxide as it would have generated if it had been burned in a gas turbine. What can be claimed for the fuel cell is that its high efficiency means that less carbon dioxide is produced for each kilowatt-hour of electricity generated than would be the case for a lower-efficiency process.

Types of fuel cell

There are six principal types of fuel cell currently under development of which four are useful for power generation applications. Their typical characteristics are shown in Table 7.1. The alkaline fuel cell, developed by Francis Bacon in the 1930s has been used in to provide power for US space ships including the space shuttle. It is extremely efficient but uses pure platinum electrodes making it too expensive for earth-bound power generation. A second type of cell which burns methanol rather than hydrogen is at an early experimental stage. Neither of these will be discussed further.

The remaining four fuel cells are all being developed for a variety of uses including power generation. Of these the phosphoric acid fuel cell (PAFC) was the first, with commercial units appearing in 1992. However these have proved expensive and this type of cell may be superseded by newer designs.

The proton-exchange membrane (PEM) fuel cell has attracted considerable attention from automotive manufacturers as an electric power source to replace the internal combustion engine. This has provided the primary stimulus for its development but there are a number of stationary power applications being developed too as well as portable applications for computers.

Both the phosphoric acid and the PEM fuel cells operate at relatively low temperatures, so they require expensive catalysts to assist the fuel cell reaction to proceed at a usable rate. The two other types under development are both high-temperature devices that do not need special catalysts. The molten carbonate fuel cell (MCFC) uses a molten carbonate electrolyte which must be heated to around 650°C while the solid oxide fuel cell (SOFC) uses a solid electrolyte that will only operate effectively at around 1000°C. The latter may eventually prove the most competitive of all fuel cell systems.

Phosphoric acid fuel cell

The PAFC uses an electrolyte composed of pure phosphoric acid, H_3PO_4. This acid is a relatively poor conductor of hydrogen ions but is stable up to 200°C and can form the basis of a fuel cell which operates using hydrogen and oxygen supplied as air.

Table 7.1 *Fuel cell characteristics*

	Operating temperature (°C)	*Catalyst*	*Efficiency (%)*
Alkaline cell	150–200	Platinum	70
PAFC	150–200	Platinum based	35–42
PEM	80	Platinum based	42–60
MCFC	650	None needed	50–60
SOFC	750–1000	None needed	50–60
Direct methanol	90–95	Platinum–ruthenium	–

The electrolyte is contained in a silicon carbide matrix where it is held within the pores of the material by capillary action. Since the electrolyte is a liquid, care must be taken to control evaporation or migration as this will impair the operation of the cell.

Electrodes made from porous carbon are bonded to the electrolyte-containing matrix using teflon mounts and the carbon is covered with a fine coating of platinum. There are groves in the back of the electrodes which carry the hydrogen or oxygen to each cell. Carbon is a good electrical conductor, so it can be used to transport the current from the cell. Each cell produces a voltage of around 0.65 V. Cells can be connected back-to-back, in series, to build what is known as a *stack*. Stacks are then connected in parallel to provide the required current and voltage output.

The PAFC requires hydrogen and oxygen, the latter from air. The cell must be heated to around 200°C before it can be started but once it starts operating the cell reaction produces sufficient heat to maintain its temperature. Water generated by the cell reaction is swept away from the cell in the air stream feeding oxygen to the cathode.

At the cell operating temperature the hydrogen–oxygen reaction will proceed sufficiently swiftly with a platinum catalyst to sustain cell operation. However the concentration of carbon monoxide in the hydrogen must be kept below 1.5% to prevent catalyst poisoning.

Hydrogen for cell operation is normally provided by reforming natural gas, though other sources such as biogas have been employed. Waste heat generated by the cell during operation can be used to carry out this reaction. The remainder of the energy which emerges as heat can be used to heat water or for space heating. The cell has a theoretical efficiency of around 42% but practical cells have not achieved this. Configured as a co-generation unit, a PAFC unit can achieve up to 87% efficiency.

PAFCs were the first type of fuel cell to achieve commercial status. A number of prototype units were field tested during the 1980s and the beginning of the 1990s. The most successful programme was carried out by International Fuel Cells[3] which produced a commercial 200-kW unit, launched in 1992. Since then over 200 have been installed across the world.

The 200-kW unit has an electrical efficiency of 36%. In addition to generating power it can supply 200 kW of hot water. As a result of its low emissions and low noise, the unit is suitable for use in urban areas and can compete with engine-based co-generation system for small commercial applications. It has the advantage of producing a greater proportion of electricity than the typical engine-based package co-generation unit.

Although these units are commercially available, they have proved costly, with a unit price in 2003 of around $900,000.[4] As a consequence, manufacture of them is likely to be discontinued and the company expects to switch to a PEM-based system for future stationary applications. Some development of the PAFC continues in Japan. However the technology may eventually be abandoned for one of the technologies below.

Proton-exchange membrane fuel cell

The PEM fuel cell uses a polymer membrane as its electrolyte. The cell was invented by US company General Electric and tested for US military use in the early 1960s. After development for the US Navy it was adopted by the British Navy in the early 1980s. Since then it has attracted most attention as a possible replacement for the internal combustion engine for automotive applications. However a number of companies are also developing stationary power applications.

The membrane which forms the electrolyte of the PEM cell is usually a compound called *poly-perfluorocarbon sulphonate*. This is a close relative of teflon but with acidic sulphonate molecular groups attached to its backbone to provide conductivity. In its normal state the membrane is not conductive but if it is allowed to become saturated with water it will conduct hydrogen ions. The membrane itself is usually between 50 microns and 175 microns thick, the latter equivalent to seven sheets of paper.

Electrodes of porous carbon containing platinum can be printed onto the membrane. A further porous carbon backing layer provides structural strength to each cell as well as supplying electrical connections. As with the PAFC, cells are joined in series and in parallel to provide sufficient current and voltage.

Since the cell contains water it must be maintained below the liquid's boiling point. Practical cells operate at around 80°C. The cell requires

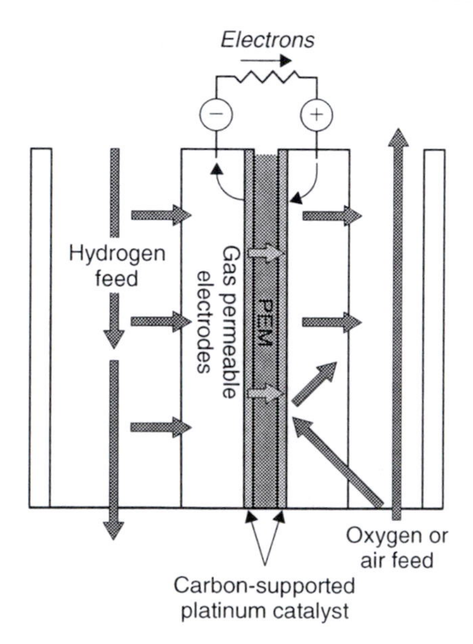

Figure 7.2 *Diagram of a PEM fuel cell*

hydrogen and oxygen (from air) to operate. As a consequence of the very low operating temperature, the hydrogen fuel must be virtually free of carbon monoxide to prevent catalyst poisoning.

The PEM operating on hydrogen has a fuel to electricity conversion efficiency of 60%, significantly higher than the PAFC cell. This, combined with the light structure makes the device attractive for automotive applications. In practice, however, the fuel is normally derived from natural gas. The low cell temperature means there is no waste heat suitable for driving the methane-reforming reaction so some fuel must be burned to provide additional heat. This reduces the overall efficiency of the cell to around 42%, similar to that of the PAFC fuel cell.

The development of PEM fuel cells has advanced rapidly as a result of investment from the automotive industry. A Canadian company, Ballard Power Systems, has built a 250-kW stationary fuel cell based on PEM technology but this prototype may never be developed commercially. Several other companies, including General Electric, and Nuvera Fuel Cells are planning units for power generation. A number of much smaller PEM units are also being developed for domestic and light commercial power applications. The first practical applications may be as a portable power source for laptop computers.

Operating a PEM fuel cell on reformed natural gas limits its performance, not only by reducing overall efficiency but also by hindering response time. While a PEM fuel cell fuelled with hydrogen can be brought on line rapidly, the reforming system is likely to require 20 min to reach operating temperature. However the technology appears to be inherently cheap. This will be the key to its success in the near term.

Molten carbonate fuel cells

The MCFC has an electrolyte which is composed of a mixture of carbonate salts.[5] These are solid at room temperature but at the cell operating temperature of 650°C, they have become liquid.

Work on high-temperature MCFCs began during the 1950s. The US army tested cells during the 1960s and in the 1970s the US Department of Energy began to support research. Japanese companies also picked up the technology. By the end of the 1990s pilot units of up to 250 kW had been tested, mainly in the USA. Development accelerated in the early years of the twenty-first century and in 2003 there were more new MCFC fuel cell installations for stationary power applications than any other technology.[6] Most, however, are still pilot or demonstration units. If these prove durable then commercial units should become available before 2010.

The MCFC has the most complex fuel cell reaction of all the cells under development. The electrolyte is a mixture of alkali metal carbonates which, when heated above 650°C, become molten and capable of conducting

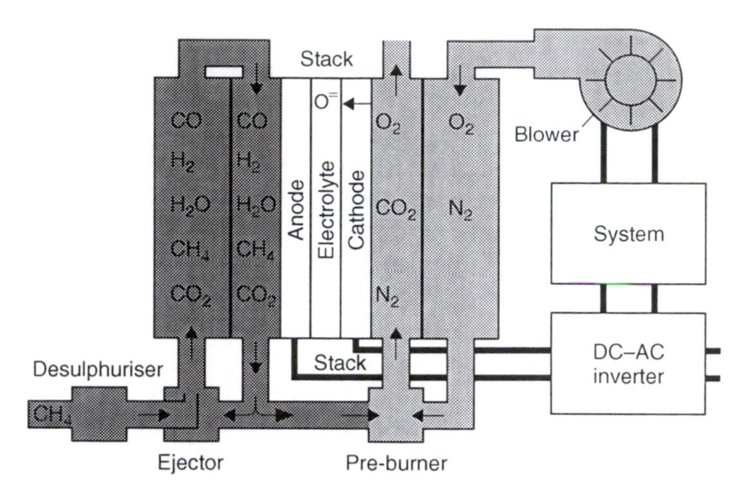

Figure 7.3 *Block diagram of a MCFC*

carbonate ions (CO_3^{2-}). Hydrogen is fed to the cell anode where it reacts with these carbonate ions, producing water, carbon dioxide and electrons. As a result of this reaction a part of the electrolyte is consumed and so must be replaced. This takes place at the cathode.

The carbon dioxide generated at the anode is carried around to the cell cathode. Here it is mixed with air. Oxygen in the air reacts with the carbon dioxide at the electrode, producing more carbonate ions which re-enter the electrolyte. In this way the composition of the electrolyte is maintained and the overall cell reaction remains that of hydrogen and oxygen reacting to produce water.

As a consequence of the elevated temperature of operation, the MCFC does not require a platinum catalyst so electrodes can be made of nickel. The electrolyte is contained in a porous refractory tile and the nickel electrodes are applied to its surface. A further advantage of the high temperature is to render the cell insensitive to carbon monoxide. In fact any carbon monoxide present will also react at the cell cathode, behaving as additional fuel.

The cell operating temperature is so high that the natural gas reformer can be built into the fuel cell itself, simplifying the system design. However the use of a high-temperature liquid electrolyte presents some significant technical challenges. Overcoming these successfully will be the key to the success of this type of fuel cell.

The reason why such a complex cell has proved worth developing lies in the potential efficiency. The theoretical conversion efficiency is 60% and though production units may only achieve around 54%, this is significantly higher than most heat-engine-based system. That does not represent the limit, however. A MCFC fuel cell produces high-temperature waste heat and this can be exploited in a gas turbine, without the need for additional fuel, to generate more electricity. In this configuration the

MCFC fuel cell may be theoretically capable of between 75% and 80% overall efficiency.

The largest MCFC power plant to be built was a 2-MW demonstration unit in Santa Clara, California. Although this plant did not achieve its design goals it provided a test bed for MCFC technology and provided the US Department of Energy with the confidence to support the development of commercial units. This was carried out in conjunction with a company called Fuel Cell Energy (FCE) which is developing 300 kW, 1.5 and 3 MW commercial units.

FCE technology is also being used by the German company MTU Freidrichshafen GmbH as the basis for a 250-kW packaged fuel cell power plant called a Hot Module. Meanwhile FCE is working with the US Department of Energy to develop a 250-kW fuel cell unit with an integrated gas turbine. This is expected to form the basis for a 40 MW power plant with an efficiency approaching 75%. Japanese and Italian companies are also developing MCFC modules.

The MCFC technology is perhaps the most complex of all fuel cell power plant technologies under development. The high-temperature operation combined with the use of a liquid electrolyte makes the technology expensive at a small scale and a unit size of 200–250 kW appears the minimum. If costs can be brought down, this technology offers the highest efficiency of all fuel cells, perhaps the highest achievable by any fuel to electricity conversion technology.

Solid oxide fuel cells

The SOFC is a very-high-temperature fuel cell. Its electrolyte is usually made from zirconium oxide, ZrO_2, zirconia. When traces of other oxides such as yttrium, calcium or magnesium oxide are added to the zirconia, it becomes capable of conducting oxygen ions. However this conductivity only becomes significant at very high temperatures and so the cell must operate at around 1000°C.

Solid oxide electrolytes were first studied during the 1930s, with little success. However work continued during the 1950s and the 1960s. The most persistent programme was carried out by the US company Westinghouse (now owned by Siemens) in conjunction with the US Department of Energy. This finally established the SOFC as a viable proposition.

As will all the other fuel cells discussed here, the cell reaction in the SOFC involves hydrogen and oxygen producing water. The difference between the SOFC and the low-temperature cells discussed earlier with acidic electrolytes is that while those provided hydrogen ion conduction, the SOFC electrolyte conducts oxygen ions. Oxygen delivered to the cathode of the cell reacts to produce oxygen ions which migrate through the electrolyte to the anode and react with hydrogen to produce water. At the elevated operating temperature this water is produced as vapour which is swept away in the fuel gas stream.

The electrolyte used in the SOFC may be as thin as 10 microns. Electrodes must be bonded to this and these also serve as a support structure to give the cell strength. Since the cell operating temperature is so high, the different materials used in cell construction must be carefully designed to have the same coefficient of expansion otherwise the cell would crack apart as it was heated.

The high temperature means that no catalyst is necessary to generate hydrogen and oxygen atoms at the electrodes. The anode is normally made from nickel dispersed in a ceramic matrix and the cathode from a conductive oxide that will not react with oxygen. The high temperature also allows the reformation of natural gas into hydrogen to take place directly on the nickel cathode.

One of the key problems with a high-temperature SOFC fuel cell is to devise a means of keeping the hydrogen and oxygen separated. The solution to this devised by Westinghouse was to build the cell as a ceramic tube with the cathode inside and the anode outside. Other designs have employed planar cells but this necessitates more complex gas routing.

The SOFC has a practical efficiency of around 50%. Its theoretical efficiency is somewhat higher though lower than that of the MCFC because of its higher temperature of operation. The simplicity of the SOFC design and its complete absence of liquids means that the SOFC fuel cell should have an extremely long operating life. Units have been tested for 60,000 h without failure and operating lives of 20 years or more can be expected.

The robust nature of the SOFC makes a hybrid design attractive too. The SOFC cell can be operated at elevated pressure, with the high-temperature exhaust gas being fed into a gas turbine to generate additional electricity. Even more power could then be derived from the system by using the waste heat from the gas turbine to generate steam in a waste-heat boiler for driving a steam turbine. Such an arrangement should be able to achieve 70% fuel to electricity conversion efficiency, possibly higher.

A large number of companies are now developing SOFC systems for power generation applications. Siemens (formerly Westinghouse) has tested a 100-kW unit and is developing both a 250-kW unit operating at atmospheric pressure and a 500-kW pressurised unit. Rolls Royce in the UK, Ceramic Fuel Cells in Australia and a variety of Japanese companies are also developing SOFC technology.

SOFC technology can also lend itself to co-generation. This has been exploited by the Swiss company Sulzer which has developed a small SOFC-based co-generation system aimed at the domestic market. The company has been testing a 1-kW unit which can also provide a household with 2.5 KW of heat.

Whatever the scale, price will be the overriding consideration with SOFC technology. The solid construction of the SOFC will lend itself to

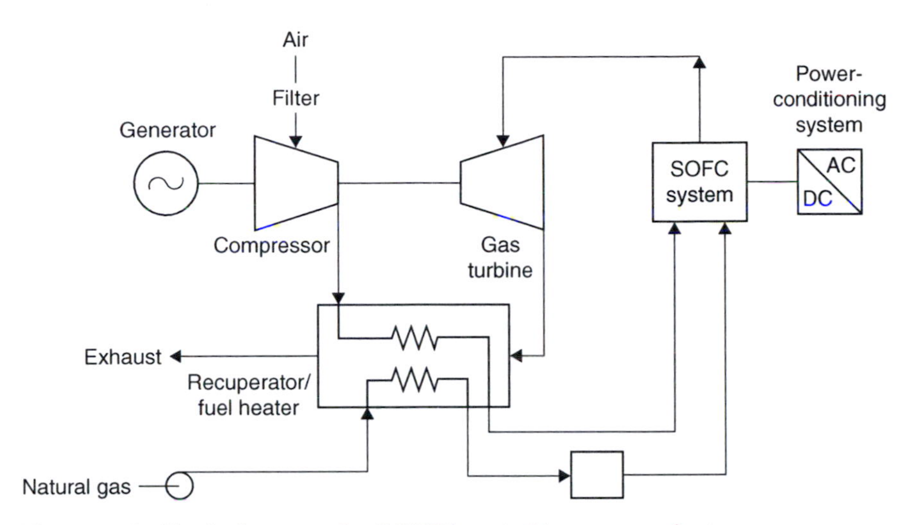

Figure 7.4 *Block diagram of a SOFC/gas turbine power plant*

simple mass production techniques which should eventually yield economies of scale. It is unlikely that commercial SOFC power plants[7] will be available at a competitive price before the second decade of the twenty-first century. Even so this is already viewed as the most promising of all fuel cell technologies.

Environmental considerations

The fuel cell is designed to consume hydrogen as fuel. A fuel cell using hydrogen has very little environmental impact since the product of the reaction is water, and water alone. The overall environmental effect of the fuel cell, therefore, depends upon the source of the hydrogen. Today most is derived from natural gas, but it could also be generated by gasification of coal or from various biomass sources. The environmental impact of each of these technologies then becomes the determining factor.

Even so, fuel cells are probably the most benign of all the power generation technologies to use fossil fuels. The levels of both sulphur and nitrogen oxide emissions they produce are extremely low, as are the particulate emissions. They are also quiet compared to the rotating machines which are normally used to generate electricity. Unit efficiency is generally higher than the efficiency of the equivalent rotating machine, so carbon dioxide emissions are proportionally lower too.

Fuel cells offer high efficiency independent of their size. A 250-kW unit will operate at exactly the same efficiency as a 200-MW power station. They also have good part-load efficiency. This makes them particularly attractive for distributed generation applications since most heat-engine-based small generating systems are less efficient than their larger relatives.

Fuel cells can easily be installed in urban areas, where population densities are high. Small, efficient units can be placed adjacent to buildings where waste heat generated by the plant can be used for heating and hot water. The water from a fuel cell can even be used as drinking water, such is its purity.

In the future it may be possible to generate hydrogen directly using renewable energy technologies. This will form the basis of a fuel economy in which hydrogen is employed as the primary means of storing and transporting energy. The fuel cell will form a key component of such a hydrogen economy, should this evolve. Fuel cells can turn hydrogen into electricity extremely efficiently, usually more efficiently than they can exploit natural gas. This will place them at the forefront of generation technologies if hydrogen becomes the fuel of choice later this century.

Financial risks

With the exception of the PAFC, fuel cells are unproven commercially. The PAFC cell has been tested in a commercial environment and has generally proved reliable. Where cost is not an overriding factor, but low environmental impact is important, then the PAFC offers a proven technology. However it is expensive (see below).

The other three technologies discussed in detail above are all unproven commercially. While demonstration projects have shown that all three technologies are viable, long-term operational experience is lacking. This should become available during incoming years so that by the end of the first decade of the twenty-first century there should be a good body of operational experience. Until that experience is available, all three technologies should be considered medium to high risk.

Fuel cell costs

The only fuel cell system for which commercial costs are available is the PAFC. Commercial 200-kW units are available for a cost of around $900,000, or $4500/kW. In practice most of these units have been installed in the USA where they have often qualified for a government subsidy of $200,000, reducing installed cost to $3500/kW. Even so, this is a high price compared to other types of small power generation or small co-generation systems.

While there may be opportunities to bring down the cost of PAFC fuel cells further, it seems unlikely that they will ever be able to achieve the near-term industry cost target of $1500/kW or the long-term target of

$400/kW by 2015. So while this first generation fuel cell is providing operating experience with the technology, it seems almost certain to be superseded by one of the other types being developed.

The PEM fuel cell is benefiting from investment from the automotive industry and this has allowed the technology to advance rapidly. The technology is almost ready for deployment and early indications suggest that a near-term installed cost of $1400/kW is achievable. General Electric believes it can market 75 kW units to the automotive industry for $500/kW by 2005[8] and $50/kW by 2010. Power production costs from first generation PEM fuel cell systems of $0.10/kWh have been suggested. This will need to be proved in service, but the near-term future of the PEM fuel cell looks promising from an economic perspective.

The MCFC has reached the demonstration stage. Early units are reputed to cost around $10,000/kW[9] but this should fall before commercial units are offered. No prices are available for demonstration SOFC units but a similar cost to that of the MCFC seems likely.

Operating and maintenance costs for all types of fuel cell should be relatively low. PAFC fuel cell stacks require replacement after around 5 years but other types of cell, particularly the SOFC, should prove more durable. The efficiency of the fuel cell and its good environmental performance should make electricity generated by fuel cell power plants attractive. They will only prove economic, however, if they can compete effectively on installed cost and that means breaking the $1500/kW barrier in the near term.

End notes

1 A heat engine becomes more efficient the hotter it is run.
2 This efficiency is achieved by the alkaline fuel cell burning hydrogen. The theoretical maximum efficiency for a fuel cell using hydrogen is 83%.
3 International Fuel Cells is a joint venture between United Technologies and Toshiba.
4 Fuel Cell Market Survey: Large Stationary Applications, Mark Cropper, Fuel Cell Today, September 2003.
5 The electrolyte is usually a mixture of sodium carbonate, potassium carbonate and lithium carbonate.
6 Refer *supra* note 4.
7 The domestic units should be available before this.
8 Refer *supra* note 4.
9 Refer *supra* note 4.

8 Hydropower

Hydropower is the oldest and probably the most underrated renewable energy resource in the world. The earliest known reference is found in a Greek poem of 85 BC. At the end of 1999 hydropower provided 2650 TWh of electricity, 19% of total global output.[1] Yet when renewable energy is discussed, hydropower barely earns a mention.

Part of the reason for this lies in the disapprobation that large hydropower has attracted over the past 10–15 years. Concern for the environmental effects of large projects which destroy wildlife habitats, displace indigenous peoples and upset sensitive downstream ecologies coupled with often heavy handed and insensitive planning and approval procedures have resulted in the image of hydropower becoming extremely tarnished.

Some of this criticism is deserved. Large hydropower projects have been built around the world without due account being taken of their effects. Schemes are often completed late and over budget. And when they are completed they sometimes do not function as intended.

This, however, is not the whole story. There are many hydropower projects that perform well. With proper planning, environmental effects can be mitigated. When accounted for properly, hydropower is one of the cheapest sources of electricity. And while the countries in Western Europe and North America have developed most of their best hydropower sites in a manner that attracts relatively little criticism today the developing world has an enormous hydropower potential which remains untapped and which, if developed sensitively, could provide a major improvement in the quality of life. Yet it is often western activists that would prevent the development of these resources.

The World Commission on Dams has addressed these problems in 'Dams and Development, a new framework for decision making'.[2] This report proposes a complete reassessment of the criteria and methods used to determine whether a large hydropower project should be constructed. It lays out an approach to decision-making which takes account of all the environmental and human rights issues which critics have raised, an approach which should filter out bad projects but allow well-conceived projects to proceed.

Large dams, however, form only part of hydropower. Small hydropower, which is generally defined as projects with generating capacities below 10 MW, can also provide a valuable source of electricity. Small projects are often suited to remote regions where grid power is impossible to deliver.

They too can have detrimental environmental effects but well-designed schemes should have little or no impact.

While large projects have been banished from the renewable arena, small hydropower is still allowed through the door. This division is politically motivated and not particularly logical since both large and small hydropower are renewable sources of energy. If the World Commission on Dams proposals are implemented then perhaps the image of large hydropower can be rescued. But that it likely to take several years, at least.

The hydropower resource

Table 8.1 presents figures for global hydropower potential, broken down by region. The gross theoretical capability figures, shown in column one, represent the amount of electricity that could be generated if the total amount of rain that falls over a region could be used to generate power at sea level (thus utilising the maximum head of water and extracting the most energy). This figure is of little practical use but the second column in Table 8.1 is more useful. This shows how much of the theoretical capability could be exploited using technology available today.

As the table shows, hydropower potential is to be found in all parts of the world. While every region has a significant resource, the largest capability exists in Asia where there is 4875 TWh of technically exploitable capability. At the other end of the scale, the Middle East has 218 TWh.

Not all the technically exploitable capability in any region can be cost effectively utilised. That which can is termed the economically exploitable capability. Of the total technically exploitable capability shown in Table 8.1, 14,379 TWh, just over 8000 TWh is considered to be economically exploitable. This is three times the 2650 TWh of electricity generated by the hydropower

Table 8.1 *Regional hydropower potential*

	Gross theoretical capability (TWh/year)	*Technically exploitable capability (TWh/year)*
Africa	>3876	>1888
North America	6818	>1668
South America	6891	>2792
Asia	16,443	>4875
Europe	5392	>2706
Middle East	688	<218
Oceania	596	>232
Total	>40,704	>14,379

Source: World Energy Council.

Table 8.2 *Regional installed hydropower capacity*

	Capacity (MW)
Africa	20,170
North America	160,133
South America	106,277
Asia	174,076
Europe	214,368
Middle East	4185
Oceania	13,231
Total	692,420

Source: World Energy Council.

plants operating around the world today. Thus two-thirds of the global resource remains unexploited.

The actual level of exploitation varies widely from region to region. The World Energy Council estimated in the 1990s that 65% of the economically feasible hydropower potential has been developed in Europe and 55% in North America. In Asia, by contrast the level of exploitation was 18% while in Africa it was only 6%.

So, as already noted, the developed world has taken advantage of much of its hydropower resource while the resource in the developing world remains largely unexploited. Africa, in particular, has some major hydropower sites that could, sensitively developed, provide significantly greater prosperity to regions of that continent.

Today the gross global installed hydropower capacity is just under 700 GW, with another 100 GW under construction.[3] Current global hydropower capacity is broken down by region in Table 8.2. In gross terms, Europe has the biggest installed capacity, followed by Asia and North America. The Middle East, probably the world's most arid region, has the smallest capacity. Comparing the numbers in Table 8.2 with those in Table 8.1 confirms that Africa has exploited relatively less of its capability than any other region.

If all the remaining economically exploitable capacity in the world was utilised with the same efficiency as that of current capacity, an additional 1400 GW could be constructed. This would roughly triple the existing hydropower capacity. Exploitation would involve an additional 14,000 power plants with an average size of 100 MW, at a cost of $1500 billion.

Hydro sites

The first stage in building a hydropower plant is to find a suitable site. This may appear obvious, but it is important to realise that hydropower is site specific. Not only does it depend on a suitable site being available but the nature of the project will depend on the topography of the site. You cannot have a hydropower plant without a suitable place to construct it. In the case of large hydro projects (>10 MW in capacity), sites will often be a long way from the place where the power is to be used, necessitating a major transmission project too.

A hydropower project requires a river. The energy that can be taken from the river will depend on two factors, the volume of water flowing and the drop in riverbed level, normally known as the *head of water*, that can be used. A steeply flowing river will yield more electricity than a sluggish one of similar size.

This does not mean that slow-flowing rivers are not suitable for hydropower development. They often provide sites that are cheap and easy to exploit. In contrast, steeply flowing rivers are often in inaccessible regions where exploitation is difficult.

Some sites offer the potential for the generation of thousands of megawatts of power. Probably the largest of these is on the Congo river where a multiple barrage development capable of supporting up to 35,000 MW could be installed. This is exceptionally large; most are smaller. Even so, such sites are likely to be extremely expensive to develop and in the current climate, extremely sensitive. They are also likely to be multipurpose projects involving flood control, irrigation, fisheries and recreational usage as well as electricity generation.

How does one set about locating a hydropower site? Many countries have carried out at least cursory surveys of the hydropower potential within their territory and provisional details of suitable sites are available from the water or power ministries. Sometimes much more detailed information is available but this cannot replace an on-site survey. Indeed surveys carried out as part of a feasibility study form a integral of any hydropower scheme.

Dams and barrages

Once a site for a hydropower scheme has been identified, there are normally two ways of exploiting it. The first is to build a dam and create a reservoir behind it from which water is taken to drive hydraulic turbines in the project's powerhouse. The second, called a *run-of-river scheme*, does without a reservoir, though it will usually involve some sort or barrage. Instead it takes water directly from the river to the powerhouse where the turbines are installed.

Run-of-river project

A run-of-river scheme is the simplest and cheapest hydropower project to develop. Since it requires no dam, a major constructional cost is avoided.

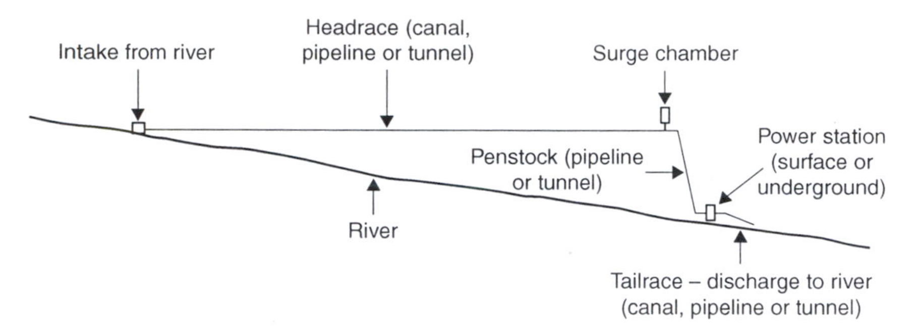

Figure 8.1 *Run-of-river hydropower scheme.* Source: *Mott MacDonald*

Geological problems associated with dam construction (see above) are avoided too. However some sort of diversion structure will be required to direct water from the river into a channel and pipework which will carry the water to the powerhouse. And if, in order to generate a significant head of water, the powerhouse is a long way from the point where the water is taken from the river (the distance can be tens of kilometres in some cases) then the geology of the route will need to be studied carefully too.

The simplicity of the run-of-river scheme is attractive but it is also the main weakness of this type of development. With no dam to conserve water, the power plant must rely exclusively on the flow of water in the river. As this fluctuates, so will the amount of power that can be generated. Under drought conditions the plant will be able to generate no power, whereas when the river is in flood, much water will have to be allowed to flow past the diversion system without being exploited. Nevertheless this type of project does have significant advantages besides cost, particularly because of the small amount of environmental disruption it causes.

Reservoir projects

The alternative to the run-of-river is the reservoir project. This will involve a major civil engineering undertaking, construction of a dam. If a dam is to be constructed, then a very careful geological survey of the underlying rock will be needed and any faults identified. Geological faults or unsuitable substrata need not prevent construction of a dam but if they are only discovered during construction they are likely to result in massive additional costs and will delay construction for months or years.

The purpose of a dam is to create a reservoir of water which builds up behind it. Once created, the reservoir allows some measure of control over the flow of water in the river beyond the dam and consequently the flow through the turbines in the powerhouse. Water can be conserved during periods of high flow and used up when rainfall is low. A dam can also be used for flood control.

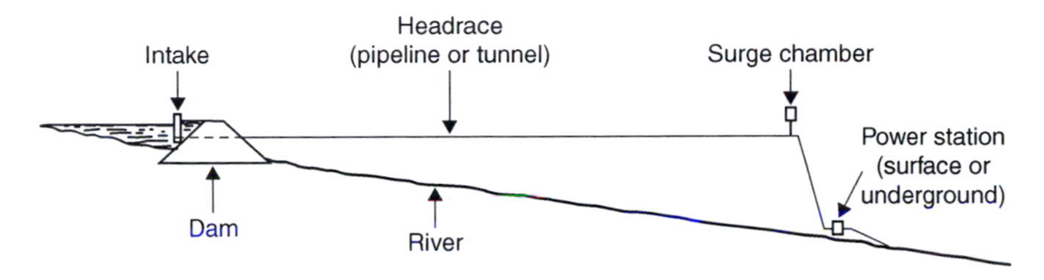

Figure 8.2 *Hydropower scheme with dam and reservoir.*
Source: *Mott MacDonald*

There are three principle types of dam used for hydropower projects, concrete dams, arch dams and embankment dams. A concrete dam is basically a massive concrete structure which, as a result of its weight, resists the pressure of water behind it. Care must be taken, however, to prevent water flowing around or beneath the dam. Concrete construction is normally employed where a high dam can be built across a narrow ravine.

Where conditions permit, the sides of such a ravine can be used as part of the construction. If the rock either side of dam site is sound and strong, it is possible to build an arch dam. In principle this acts in exactly the same way as an arch in a building, but with the bow of the arch facing upstream to resist the pressure of water behind it rather that vertically, supporting the weight of a building. Provided the sides of the arch dam are anchored securely to the rock at each side, the arch is incredibly strong and requires much less material than other types of dam.

When a broad, shallow dam is required, an embankment dam is the more normal choice. This is constructed from a mixture of materials but the major component is usually earth if this is available locally. The dam allows a certain amount of water to seep through it. This must be carefully controlled to prevent damage to the structure. It is also vital to ensure that the water in the reservoir does not flow over the top of the dam; if it does it could wash the structure away.

The cost of a dam is a major factor in the financing of a hydropower project. The dam is also the part of the project which is likely to cause the most controversy. A reservoir behind a dam will inundate a large area of land, displacing people and destroying habitats. Downstream habitats may also be effected by the reduced flow of water, at least while the dam is filling. Detailed environmental impact studies will normally be required before such a project can proceed.

Turbines

Hydropower exploits the energy contained in the water of rivers and streams to produce electricity. Dams, canals and high-pressure pipes control and transport this water but the key technological element involved in the energy conversion process is a hydraulic turbine.

The hydraulic turbine is a simple, reliable and well-understood component, made from simple materials. Most turbine are made from iron or steel. In the past wood was commonly used too.

The history of the hydro turbine is long. Water wheels for grinding grain were used by the Romans and were known in China in the first century AD. They were common across Europe by the third century AD and could be found in Japan by the seventh century. The Doomsday Book of AD 1086 records 5000 in use in the south of England. These early water wheels were made of wood. Iron was first used in the eighteenth century by an English engineer, John Smeaton.

The modern hydraulic turbine is a direct successor of Greek, Roman and Chinese machines. However development work in the nineteenth century has led to two distinct branches of turbine design. These are usually called *impulse turbines* and *reaction turbines*.

Impulse turbines

The main type of impulse turbine in use today is the Pelton turbine, patented by the American engineer Lester Allen Pelton in 1889. It is found mainly in applications where a high head of water is available. Another type, called the Turgo turbine, has also been developed, again for high-head applications. In both cases, the head of water will normally be greater than 450 m, although the Pelton turbine is applicable for heads of between 200 and 1000 m. (For heads higher than 1000 m there will probably be two turbines, each exploiting half the head.)

A high head of water will generate an enormous pressure at its base. If the water is released through a narrow nozzle, the pressure of water will generate a fierce jet of water. The impulse turbine harnesses this energy of motion.

The Pelton turbine has bucket-shaped blades. The high-pressure jet of water is directed into the buckets at an angle that ensures that the energy in the water is virtually all converted into rotary motion of the turbine wheel. This conversion process can achieve an efficiency of nearly 95%, under ideal conditions, so little energy is wasted.

One of the keys to the operation of an impulse turbine is that it must rotate in the air. If it becomes submerged, its rotation is hampered. This is in direct contrast to the second type of turbine, the reaction turbine, which must be submerged to operate efficiently.

Reaction turbines

For heads of water below 450 m, a reaction turbine will be the normal choice. This type of turbine must be completely submerged to operate

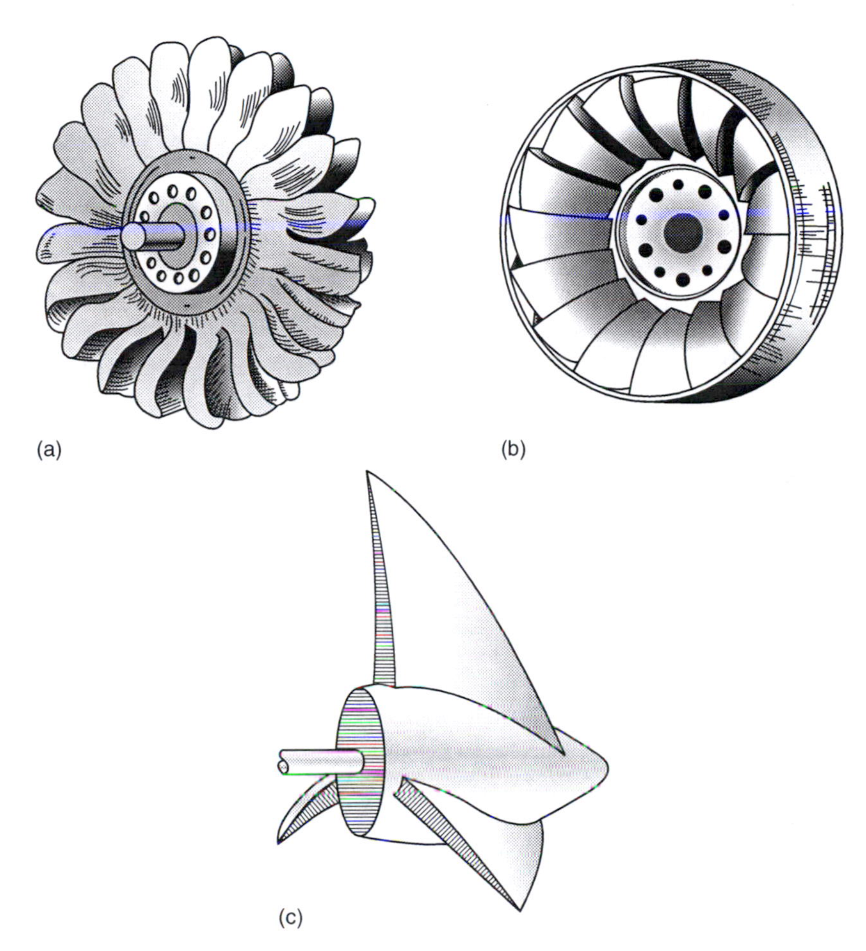

Figure 8.3 *Hydropower turbines: (a) Pelton (b) Francis and (c) propeller turbines*

efficiently. And whereas the impulse turbine harnesses the kinetic contained in a jet of high-pressure water, a reaction turbine responds to the pressure (potential energy) from the weight of water acting on one side of its blades.

There are several different types of reaction turbine. The most popular, accounting for 80% of all hydraulic turbines in operation, is the Francis turbine. This can be used in almost every situation but for very low heads, propeller turbines and Kaplan turbines are frequently preferred.

Francis turbine

The Francis turbine was developed by James Bichens Francis around 1855. Its key characteristic is the fact that water changes direction as it passes through the turbine. The flow enters the turbine in a radial direction, flowing towards its axis, but it exits along the direction of that axis. It is for this reason that the Francis turbine is sometimes called a *mixed-flow turbine*.

The blades of a Francis turbine are carefully shaped to extract the maximum amount of energy from the water flowing through it. Water should flow smoothly through the turbine for best efficiency. The force exerted by the water on the blades causes the turbine to spin and the rotation is converted into electricity by a generator. Blade shape is determined by the height of the water head available and the flow volume. Each turbine is designed for a specific set of conditions experienced at a particular site. When well designed, a Francis turbine can capture 90–95% of the energy in the water.

Francis turbines are the heavyweights of the turbine world. The largest, at the Itaipu power plant on the Brazil–Paraguay border, generate 740 MW each from a head of 120 m.

The Francis design has been used with head heights of from 3 to 600 m but it delivers its best performance between 100 and 300 m. Impulse turbines are often preferred for higher heads while lower heads are exploited using propeller and Kaplan turbines.

Propeller and Kaplan turbines

The propeller turbine looks like the screw of a ship, but its mode of operation is the reverse of the ship's propulsion unit. In a ship a motor turns the propeller which pushes against the water, forcing the ship to move. In the hydropower plant, by contrast, moving water drives the propeller turbine to generate power.

Propeller turbines are most useful for low-head applications such as slow running, lowland rivers. Their efficiency drops off rapidly when the water flow drops below 75% of the design rating so plant designers often use multiple propeller turbines in parallel, shutting down some when the water flow drops in order to keep the remaining turbines operating at their optimum efficiency.

In some cases multiple turbines will be inappropriate, even though flows are not steady. Under these circumstances, a single turbine can provide better performance under variable flow conditions if the angle of the blade on the turbine can be varied. This is the principle of the Kaplan turbine.

Another variant is the bulb turbine, used for extremely low-head conditions. In this design the turbine and a watertight generator are enclosed in a bulb-shaped container. The turbine rotor can have fixed or variable blades. Water flows into one end of the bulb-shaped container and out the other, with no change of direction. The bulb turbine has been used in tidal power plants.

Generators

Most hydropower plants employ conventional generators with one generator for each turbine. Generally both the turbines and the generators are designed for a specific site and the turbine and generator speeds are fixed. More recently variable-speed generators have also started to appear in hydropower applications. These allow an additional degree of flexibility by allowing the turbine speed to be varied in order to operate at the optimum efficiency under differing flow conditions. However variable-speed generators are generally more expensive than their fixed-speed equivalents.

Small hydropower

Small hydropower projects are those under 10 MW in size, though this classification can vary from country to country. (In China, for example, any project under 25 MW is considered small.) While small projects operate on essentially the same principles as large projects and use similar components, there are differences that need to be considered separately.

There are three types of small hydropower project, designated small, mini and micro. According to a United Nations Development Programme (UNDP)/World Bank definition, a project in the range 1–100 kW is classified as a micro project while 100 kW to 1 MW is a mini project. The small project range will stretch from 1 MW to between 5 and 30 MW depending on who is defining it.

Small hydropower potential is often assessed separately from large-scale hydro potential. A 1996 estimate[4] put the global small hydro capacity at 47,000 MW with a further 180,000 MW remaining to be exploited. In Europe there is around 9000 MW of installed small hydro capacity and sites exist for 18,000 MW more. China claims an exploitable potential for sites with capacities under 25 MW of 70,000 MW. Madagascar maintains it has a gross theoretical small hydro potential of 20,000 GWh each year. Clearly there is enormous potential for future development in may corners of the world.

Small hydropower projects can be developed anywhere, but mountainous terrain often offers the best potential. Thus Austria and Switzerland are both big users of small hydropower in Europe. This represents a valuable resource since communities located in mountainous terrain often cannot be connected to a national grid.

Small hydropower plants are conceptually similar to their larger siblings but the level of investment involved will affect the way a small project is developed. The turbines used in small plants are the same types as those employed in large projects but whereas the large plants will use turbines designed specifically for the site being developed, a small plant will normally have to use off-the-shelf turbine designs and generators in order to keep costs down.

In addition to the standard Pelton, Francis and propeller turbines, there are a number of special small hydro turbines. These include Mitchell Banki turbines, Turgo impulse turbines, Osberger crossflow turbines and Gorlov turbines. Energy efficiency tends to be lower for small hydro projects. A study by the UNDP and the World Bank in Ecuador[5] found that systems under 50 kW had a maximum efficiency of 66% rising to 70% for units in the 50–500 kW range and 74% for units between 500 kW and 5 MW.

Head height is an important factor in determining small hydro economics with higher head sites generally cheaper to develop. An impulse turbine is the best choice where the head height is above 30 m, a reaction turbine below for lower heads. A head height of less than 2.5 m is difficult to exploit.

Dam and barrage structures are also similar in small and large projects but many small schemes use simpler designs. Run-of-river designs are popular since they involve the minimum of civil works. Novel designs, such as inflatable and rubber barrages have also been employed.

A key cost factor in a small hydro project is the feasibility study. Any hydro project must involve a pre-feasibility study to determine if the site is suitable for development and a feasibility study to prepare design details. The studies will look at the hydrological and geological conditions at the site. For a large scheme the feasibility study normally accounts for 1–2% of the total cost. In a small scheme it has been known to consume 50% of the budget.

Small hydro budgets are squeezed from other directions too, because capital costs do not necessarily fall in proportion to the size of the scheme. Control system costs, for example, escalate as the project size falls. The cost of grid connection may also make smaller projects uneconomical as grid-connected public power providers, although they can still provide an economic supply to a small isolated village or hamlet.

The environment

The environmental effects of a hydropower project, particularly one involving a dam and reservoir, are significant and must be taken into account when a project is under consideration. What is going to be submerged when a reservoir is created? What effect will the dam or barrage have on sedimentary flow in the river? What are the greenhouse gas implications? Who's interests are affected? All these issues must be addressed.

In order to make a case for such a project, a thorough environmental assessment will usually be necessary and in most cases it will be mandatory. Such a study should include proposals for the mitigation of any negative effects of the development. In many cases, particularly where international lending agencies are involved, a project will not be permitted to proceed unless the environmental assessment is favourable. This is equally true of public sector and private sector projects.

Inundation

After a dam has been constructed, an area of land behind it is inundated to create a storage lake. It is the loss of this land that normally leads to the greatest controversy. (Run-of-river schemes are less disruptive and may be considered acceptable in situations where a dam and storage reservoir are not permissible.)

The most significant effect of a reservoir will be to displace people living in the area to be flooded. Resettlement is extremely disruptive, particularly if the land involved has cultural and ancestral associations stretching back over decades or centuries. If a unique ethnic group is involved it is unlikely that the project will be permitted to proceed. But in all cases resettlement should be minimised.

If resettlement is permitted, human rights considerations dictate that it should be carried out in consultation with the people involved. A hydropower scheme is usually intended to improve the local standard of living and that yardstick should be applied to the displaced people; these people should be better off after displacement than they were before.

To achieve this aim involves financial support which should be built into the project budget. As a rule of thumb, a figure of six times the per capita gross national product (GNP) of the host country should be allowed for each individual to be resettled.

Effects on plant and animal life in the area must also be taken into account. Unique habitats will need replacing with new habitats in the region of the reservoir. The effect on fish, particularly migratory fish such as salmon and eels must be studied. In addition a reservoir can stimulate seismic activity as a result of the pressure of impounded water; its likelihood should be assessed.

Sedimentation

Another effect of a dam or barrage is to change the sedimentation regime in a river. Most rivers carry some sediment downstream with them. Some, such as the Nile, carry enormous quantities of fertile material upon which a whole civilisation has depended.

When a dam is built across a river the reservoir behind it reduces the flow rate of the river and much of the sediment can precipitate onto the reservoir floor. In the worst case this will lead to the eventual filling of the reservoir with sediment. In more propitious circumstances a steady state will eventually be reached and transport of sediment downstream will become reestablished.

Whatever the situation, the amount of sediment flowing past the dam will be reduced, at least initially. This can have important consequences downstream. Erosion rates may increase in the riverbed below the dam.

More seriously, important ecostructures which rely on the sediment and its nutrients may become seriously disrupted or even destroyed.

Inter-regional effects

The construction of a dam on an inland section of a major river will have a significant effect of both water flow and sediment flow below the dam. This will effect many downstream communities because it can disrupt ecosystems, affect fishing, remove water that has previously been used for agricultural purposes and change pollution levels.

These downstream effects will often be experienced in a different region of the country to that in which the dam is built. In some cases (such as the construction of dams on the river Euphrates in Turkey) it will affect conditions in a different country. Such effects are easily ignored, but modern environmental considerations dictate that this should not be allowed. All affected parties should be considered when such a project is considered and it should only be constructed with their agreement.

Greenhouse gases

While many of the effects of a hydropower project are negative, the effect on greenhouse emissions should, on the face of it, be positive. The generation of hydropower does not involve creation of carbon dioxide. Unfortunately the situation is not that simple because a reservoir can become the source of methane and this gas is an even more efficient greenhouse gas than carbon dioxide. (It is roughly eleven times more potent.)

A reservoir will become a source of methane if it contains a great deal of organic material – a tropical rain forest would be ideal – and conditions are right for anaerobic fermentation. In the worst case a hydropower plant can produce more greenhouse emissions, over its lifetime, than a similarly sized fossil-fuelled power plant.

Fortunately that is not normally the case. If the site is chosen carefully, and trees are cleared before inundation, the project should produce total greenhouse emissions equivalent to as little as 10% of the emissions in 1 year from a similarly sized fossil fuel plant. Most of that will be carbon dioxide generated as a result of the construction of the components of the plant.

Human rights

As already indicated above, the construction of a major hydropower project will often affect the lives of a wide range of people. These people generally have rights over land that will be used or rights of water use. Under the principles outlined by the World Commission on Dams[6] each group or individual with an interest should be consulted when such a project is proposed and the project should not normally proceed without agreement from all the parties.

The World Commission on Dams report also suggests that such considerations should be applied to existing dams. There are many cases around the world where groups have been affected by dam construction but were never properly consulted. That they should be considered now is an ambitious proposal which is unlikely to be welcomed by governments in many parts of the world.

Financial risks

Hydropower relies on running water to generate electricity. This water is provided by the rain cycle, a natural process outside the bounds of human control. Consequently it is impossible to guarantee the output from a hydropower plant at any given time in the future. Nevertheless the output can be guaranteed with a fair degree of certainly over a long time scale; the greater the period, the more certain the predictions will be.

This hydrological risk – the risk that there may be periods with no water in the river where a plant is operating – can be quantified in the same way as the risks involved in other types of power plant project. In fact one could argue that it can be more precisely quantified than the risk associated with, for example a fossil fuel supply, where the supply chain depends on human intervention.

The second major risk associated with hydropower is geological risk. Geology is seen as a problem because too many developers in the past have not taken geological factors into full account. It is wise to assume that every hydropower project will face some geological problem that will complicate its construction. The complication-free project is the exception. This may appear to be a bleak prognosis, but it is pragmatic. Once the prognosis is accepted and factored into the project, cost and construction overruns become manageable.

Geological risk

Geological risk is associated with the geology of the site upon which the project is to be built. What type of rock lies beneath the construction site? Are there fault lines running through it? Is the area subject to seismic activity? The answers to these, and other, questions affect both the economic viability of a scheme and its eventual design.

There is only one way to discover the answer to these questions, and that is by carrying out a thorough feasibility study of the site. Such a study will be expensive, but it is vital if the project is to be understood and managed successfully. If the study is not carried out, and unexpected geological features are discovered during construction, cost and time overruns can be enormous.

Virtually all hydropower schemes have to face some unexpected geological problem. Most can be overcome. But knowing what is coming is the only way to manage the finances and construction timetable with any degree of certainty. Even then the actual construction may throw up some new difficulties.

The projects which have suffered most in this respect are those where the underground structure was thought to be known and no project-specific study was carried out. The lesson is clear. A thorough feasibility study is imperative. Understanding the geology of the site is essential if the project is to be designed correctly.

The possibility of seismic activity must also be taken into account when designing a hydro project. If earth tremors are likely, the dam and powerhouse must be designed to withstand them. A major earthquake during construction could also cause enormous disruption, particularly to temporary diversion structures. There may be cases where it would be imprudent to build a dam because of the risk of earthquake. In many cases, however, good design will enable a dam to withstand an earthquake without damage.

Hydrological risk

The feasibility study must also analyse carefully the hydrology of the project site. This involves establishing the expected flow in the river that is being harnessed for the project.

The only sure way of determining flows is from historical records. Some countries, particularly, European countries and countries that were once colonies of European powers, will often have good hydrological records. In other areas the records will be non-existent.

If there are no records, then it is possible to reconstruct them by secondary means but this will never be as reliable as an accurate set of records. Opinions differ about the length of record needed but 10 years is normally thought too short a period and 40 years barely sufficient.

The historical record will show how river flows have varied. It will indicate the maximum and minimum flows to be expected and the average flows. These figures will not allow prediction of the amount of water in the river at any specific date in the future but they will allow average generation levels to be computed. It must, nevertheless, be borne in mind that there will be days and months when flow is minimal or none existent. And there will always be a risk of flood.

Hydrological records provide the data upon which to base a power purchase agreement for a hydropower plant. However the records cannot take account of one factor, future upstream use of water.

When Turkey built the Ataturk dam on the Euphrates, flow through Syria and Iraq was seriously affected. In fact it stopped completely for a month in 1990 when the reservoir was filled. This is a risk that cannot be tackled at a project level. Risks of this sort are extremely difficult to predict. If upstream development does occur, then legal recourse represents the only way of gaining adequate compensation.

Global climate change can also affect rainfall patterns and hence river flows. Changes are likely to take place slowly and can be predicted with careful analysis. This is another effect that should be taken into account in any feasibility study.

The cost of hydropower

As with many renewable sources of energy, most of the costs associated with a hydropower plant are up-front costs required for its construction. Under most circumstances the actual source of energy, the water, will cost nothing.

In the case of hydropower the up-front costs can be high. This can make hydropower plants difficult to fund using standard lending arrangements. Project financing in particular, where a loan is made in the expectation of payback being covered by revenue from the power plant, has proved particularly difficult in recent years. The interest payments required force the cost of electricity too high for it to be economical.

And yet, costed realistically, hydropower is certainly competitive. Some would argue that it is the cheapest sources of electricity available. The problem for hydropower is that while commercial loans for power plants are generally over 10–20 years, a hydropower plant will continue to generate power for perhaps 50 years; with relatively small further investment to rehabilitate the powerhouse, this can be extended to 100 years or longer. There are some dams still functioning in Spain that were built by the Romans – though not to generate power.

The cost of hydropower varies from country to country and project to project. Table 8.3 lists some plants built in the last two decades (the Fiji plant was actually completed in 1982). As the table shows, the cost of construction of a project can range from $700/kW to $3500/kW.

The Chinese government has invested heavily in hydropower over the last decade. Experience there indicates that medium- and large-scale projects can be built for an average cost of around $740/kW. In general smaller projects are relatively more costly, as Table 8.3 indicates. Remote sites such as those in Nepal are also more costly to develop than easily accessible sites. Project costs will also depend on the type of hydropower plant being

Table 8.3 *Typical hydropower project costs*

	Capacity (MW)	*Cost (US$ millions)*	*Unit cost (US$/kW)*
Upper Bhote Koshi (Nepal)	36	98	2722
Manasavu-Wailoa (Fiji)	40	114	2850
Kimti (Nepal)	60	140*	2333*
Bakun (Philippines)	70	147	2100
Mtera (Tanzania)	80	139	1738
Casecnan (Philippines)	140	495	3536
Theun Hinboun (Laos)	210	317	1510
San Roque (Philippines)	345	580	1681
Birecik (Turkey)	672	1236	1839
Ita (Brazil)	1450	1070	738
Karakaya (Turkey)	1800	1496	831
Three Gorges (China)	18,200	15,000*	824*

*Estimated costs.
Source: World Bank, Statkraft, Modern power systems, *The International Journal on Hydropower and Dams*, Montgomery Watson Harza.[7]

built. Turbines for low-head power plants tend to be more expensive than those for high-head projects. Bulb turbines, of any size, are inherently costly.

The cost of electricity from a hydropower plant will depend on the cost of building and financing the project and on the amount of electricity it generates when operating. For recent hydropower projects built by the private sector with loans repaid over 10–20 years, initial generation costs have been in the range $0.04–0.08/kWh.[8] However once the loan has been repaid the costs drop dramatically. The typical range of generation costs is $0.01/kWh–$0.04/kW but may easily fall below 0.01/kWh. This is cheaper than any other source of electricity.

Small hydro projects can range from $800/kW to over $6000/kW depending on the site and the size of the scheme. According to the Indian Renewable Energy Development Agency, the capital cost of small hydro in India is between $800/kW and $1300/kW and the generation cost is $0.03–0.05/kWh. Similar figures from the Energy Technology Support Unit for a typical UK project put the capital cost at around $1500/kW.

End notes

1 World Energy Council, Survey of Energy Resources, 2001.
2 Dams and Development, a new framework for decision making, The World Commission on Dams, Earthscan, 2000.

3 World Energy Council, Survey of Energy Resources, 2001.
4 European Small Hydro Association Document 96011, C. Penche, 1996.
5 Private minihydropower development study: the case of Ecuador, UNDP/World Bank, November 1992.
6 Dams and Development, a new framework for decision making, Earthscan, 2000.
7 The need for a new approach to hydropower financing, Bruno Trouille, Montgomery Watson Harza.
8 Refer *supra* note 7.

9 Tidal power

Tidal power stations take advantage of the tidal rise and fall to generate electricity. In the simplest type of tidal development, a barrage is built across the estuary of a river. When the tide rises, water flows from the sea into the estuary, passing through sluice gates in the barrage. At high tide the sluice gates are closed and when the tide ebbs, the water behind the barrage is allowed to flow back to the sea through hydraulic turbines, generating power in the process.

Exploitation of tidal motion has a long history but its use for power generation is extremely limited with only a handful of operating plants in existence in the world. Tidal electricity generation is only possible in locations where the tidal span (the distance between high and low tide) is significant. Globally, a number of favourable sites have been identified but the costs involved in building a barrage make tidal power uneconomical to develop today.

Another means of extracting energy from tidal motion is to use an underwater windmill. This technology is considered separately in Chapter 14 which is devoted to ocean power.

Tidal motion

The motion of the tides is caused primarily by the gravitational pull of the moon and the sun. This motion varies according to a number of cycles.

The main cycle is the twice daily rise and fall of the tide as the earth rotates within the gravitational field of the moon. A second, 14-day cycle which results in spring and neap tides is caused by the moon and the sun being alternately in conjunction or opposition. There are other cycles that add 6 monthly, 19- and 1600-year components but these are much smaller.

Tidal amplitude in the open ocean is around 1 m. This increases nearer land. Amplitude can be substantially enhanced by the coastal land mass and by the shape of river estuaries. Under particularly propitious conditions, such as are found in the Severn estuary in southwest England, or the Bay of Fundy in Canada, the tidal amplitude will increase substantially. For example, the Severn has an exploitable amplitude of 11 m.

The energy that can be extracted from tidal motion waxes and wanes with the tide itself. Power output is generally not continuous. It is, however, extremely predictable. Unlike most other forms of renewable energy, it is not subject to vagaries of weather. This means that the future output of a tidal power station can be determined with great accuracy.

The tidal resource

The World Energy Council has estimated the global annual energy dissipation as a result of tidal motion to be 22,000 TWh. Of this, 200 TWh is considered economically recoverable. Less than 0.6 TWh is actually converted into electricity.

There has been considerable interest in tidal power since the 1960s and a number of countries have identified sites where tidal power production would be possible. However in most cases proposed schemes have been judged too expensive to build.

One of the most thorough research projects into tidal potential was carried out in the UK between 1983 and 1994. This project looked at a range of possible schemes in England and Wales. It concluded that if every practicable tidal estuary with a spring tidal range of more than 3.5 m was exploited, around 50 TWh of power could be generated each year. This represented around 20% of the electricity consumption in England and Wales in the mid-1990s. The UK's best site is the Severn estuary.

In Canada, the Bay of Fundy has the highest tides in the world. This region, on Canada's east coast, has been the subject of intense examination. A comprehensive study of the region, carried out in the mid-1960s, focussed on sites with a total generating capacity of nearly 5000 MW. However tentative schemes to build projects were abandoned during the changing economic climate at the end of the 1970s.

Russia has significant potential for tidal generation, particularly in the White Sea on the Arctic coast and in the Sea of Okhotsk. A site at Tugur bay with the potential to generate 6800 MW has been identified as promising but the future of this project is uncertain.

Korea has potential tidal sites on the country's east coast. Consultants from various countries have carried out studies at several of these sites. Tidal span on this coast is not great but the region benefits from reflection from the South China Sea. The most promising project is at Garorim where a scheme with a projected capacity of 400 MW has been studied.

India also has substantial tidal potential. The Gulf of Kutch on the northwest coast has been studied and a 600 MW project proposed. The Indian government has estimated the country's tidal potential to be 10,000 MW.

China has studied various potential sites. Its southeast coastline is thought to offer particularly good opportunities. Mexico has looked at a site on the Colorado estuary, Brazil and Argentina have studied projects and the USA has examined a site in Alaska.

Australia's northwestern coast has some of the highest tidal ranges in the world and there are a number of inlets which could be harnessed to generate electricity. A novel two-basin project was proposed near the town of Derby, but the scheme was rejected by the Western Australian government in 2000 in favour of a fossil fuel plant. However other projects are being discussed.

Tidal power need not be tied to estuaries. In the 1960s France developed plans for an offshore project in Mont St Michel bay. The scheme was shelved when the country decided to invest heavily in nuclear power.

The Mont St Michel project involved a tidal plant that did not make use of an estuary. Instead, a circular barrage was to be constructed which would completely enclose an area of open sea. This type of plant would operate in exactly the same way as an estuary plant, with water flowing into the enclosed reservoir when the tide rises, and flowing out through turbines during the ebb tide. While this approach would involve enormous construction costs, it does have the merit of allowing a large tidal plant to be built where no suitable estuary exists.

Tidal technology

Harnessing tidal motion to generate mechanical power has a long history. Tidal basins were being used in Europe to drive mills to grind grain before AD 1100. These plants were only replaced when the Industrial Revolution introduced steam engines and fossil fuel.

The exploitation of tidal ebb and flow to generate electricity has been less well tried. Table 9.1 shows the most important tidal power plants that have been built this century. As this table indicates, the largest is La Rance on the northwest coast of France close to St Malo.

The 240 MW La Rance plant uses specially devised bulb turbines. A small turbine of similar design was bought by Russia during the 1960s, and promptly disappeared from sight. There has since been speculation that the 400 kW project Kislaya Guba represents the final resting place for this turbine.

After La Rance, the second largest project is at Annapolis Royal on the Bay of Fundy in Canada. China has also developed some small-scale projects,

Table 9.1 *The world's tidal power plants*

Site	*Country*	*Capacity (MW)*	*Year entered service*
Various	China	11.0	1958 onwards
La Rance	France	240	1966
Kislaya Guba	Russia	0.4	1968
Jiangxia	China	3.2	1980
Annapolis	Canada	17.8	1984

World Energy Council, Modern Power Systems.

of which the largest are a 3.2 MW project at Jiangxia and another 5 MW plant. Work on tidal power generation began in China in 1958 and there are thought to be seven projects in operation today with an aggregate capacity of 11 MW.

Tidal barrages

The construction of a tidal barrage represents the major cost of developing tidal power. As a result, much of the research work carried out into tidal power has focussed on the most efficient way of building the barrage.

Construction of the French tidal power plant at La Rance was carried out behind temporary coffer dams, enabling the structure to be built under dry conditions. While La Rance was completed successfully using this approach, the method is generally considered too expensive as a means of constructing a tidal barrage today. There is also an environmental problem attached to completely sealing an estuary for the period of construction, which might easily stretch into years. For that reason, such an approach is unlikely to be adopted for the future.

A novel approach suggested for the construction of a barrage across the River Mersey in England borrows something from the construction of La Rance. The idea proposed was to procure a pair of redundant bulk carriers, oil tankers for example, and sink them on the riverbed parallel to one another, sealing the ends and filling the enclosed space with sand to create an island. Concrete construction would be carried out on the island as if it were dry land. To create a watertight structure, diaphragm walls would be fabricated of reinforced concrete; the turbines and sluice gates required for the operation of the power station would subsequently be fitted to this concrete shell.

Once the first section of the barrage had been completed the bulk carriers would be refloated, moved along to the next section and sunk again.

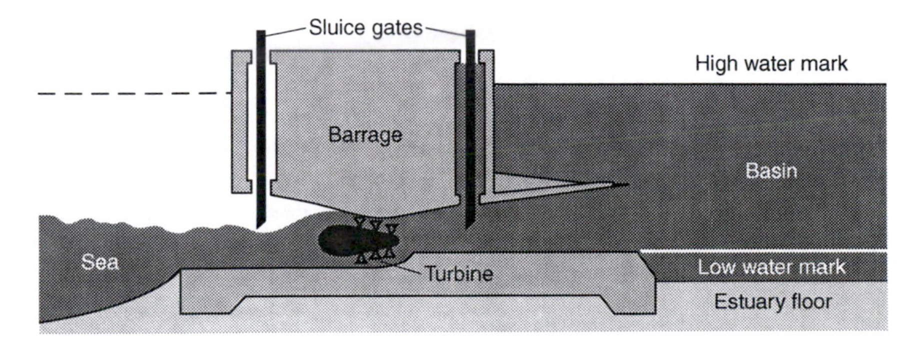

Figure 9.1 *Cross section of a typical tidal barrage*

This process would be repeated, until the barrage had been completed. The Mersey barrage has not been built, so the efficacy of the method has yet to be tested.

Where an estuary is shallow, an embankment dam could be constructed using sand and rock as its main components. Sand alone would not make a stable embankment; wave erosion would soon destroy it. Hence, some form of rock reinforcement would be required on the seaward side. Concrete faces on both sides of the embankment could provide further protection.

The sand needed for construction of such an embankment can easily be recovered from the estuary by dredging. Rock, which must generally be blasted from the riverbed, is a more expensive material and its use needs to be minimised.

While all these methods have their attractions, the construction method most likely to be used to build a large barrage today would involve prefabricated units called *caissons*. Made from steel or concrete, the caissons would be built in a shipyard and then towed to the barrage site where they would be sunk and fixed into position with rock anchors and ballast.

Some caissons would be designed to hold turbines; others would be designed as sluice gates and a third type would be blank. These would be placed between the other two types to complete the barrage.

Caisson construction was the favoured approach in a study for construction of the Severn barrage in England completed in 1989 under the auspices of the Severn Barrage Development Project. A turbine caisson for this project would have weighed over 90,000 tonnes and would have a draft of 22 m. The minimum height of the vertical faces would be 60 m. As a result of their size, special facilities would have been needed to construct them.

Prefabrication of the caissons was expected to reduce construction time to a minimum. Even so, the Severn project was scheduled to take 10 years to complete. There remained some uncertainty about how easy it would be to place the caissons in position, uncertainty that could only be dispelled by actual construction.

Two-basin projects

The simplest tidal power plant has a single basin or reservoir formed behind a barrage across a river estuary. More complex designs are possible. The most interesting of these are two-basin designs. A number of two-basin tidal power schemes have been proposed though none has yet been built. The advantage of a two-basin project is that it can generate power either continuously or for a longer period of time than in a single-basin project.

The best-developed project of this type was one proposed for construction near Derby in Western Australia. The project involved building barrages across two adjacent inlets and creating an artificial channel connecting the two basins formed by these barrages. A power station with turbines capable of generating 48 MW was to be stationed on this artificial channel.

Operation of the plant would involve maintaining a high water level in one basin and a low water level in the second. This would be achieved by opening sluice gates in the first barrage when the tide was at its high point and opening gates in the second barrage a low tide. Provided the basins were generously enough sized, water could flow continuously through the artificial channel from the high to the low basin without reducing the head of water between the two basins significantly. Thus power could be generated continuously. Other variations on this scheme are possible.

Bunded reservoir

As an alternative to a barrage across an estuary, it is theoretically possible to enclose an area of a tidal estuary or tidal region of the sea with an embankment or bund (see the St Malo project discussed above). The principle involved is the same, creating a reservoir that can be filled at high tide and then allowed to empty when the tide has fallen. The environmental effect of such a structure would probably be less dramatic than complete closure of an estuary but costs are likely to be higher.

Turbines

The turbines in a tidal power station must operate under a variable, low head of water. The highest global tidal reach, in the Bay of Fundy in Canada, is 15.8 m; most plants would have to operate with much lower heads than this.

Such low heads necessitate the use of a propeller turbine, the turbine type best suited for low-head operation. The fact that the head varies appreciably during the tidal cycle means that a fixed-blade turbine will not be operating under its most efficient conditions during the majority of the tidal flow; consequently a variable-blade Kaplan turbine is usually employed.

The most compact design of propeller turbine for low-head applications is the bulb turbine in which the generator attached to the turbine shaft is housed in a watertight pod, or nacelle, directly behind the turbine runner. The La Rance tidal plant employs 24 bulb turbines, each fitted with a Kaplan runner and a 10 MW generator.

Bulb turbines were new when La Rance was built and construction of the plant involved some experimental work; of the 24 turbines, 12 had steel runners and 12 had aluminium bronze runners. Experience has led the operators to prefer the steel variety.

The turbines at La Rance were designed to pump water from the sea into the reservoir behind the barrage at high tide to increase efficiency. This

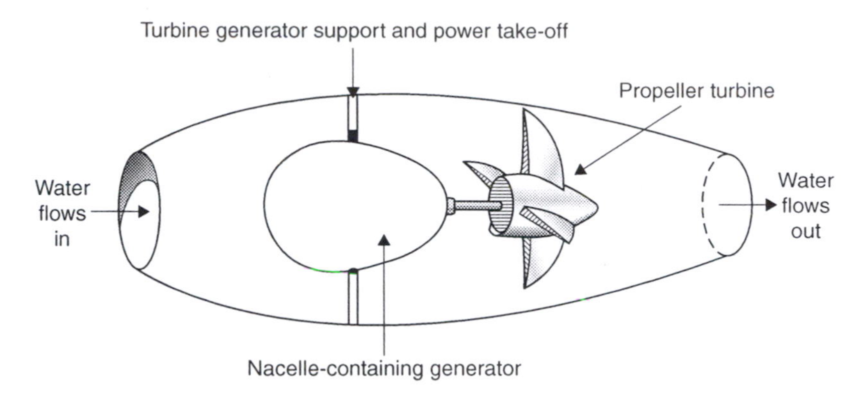

Figure 9.2 *Bulb turbine*

was found to cause severe strain on parts of the generator and the design had to be modified. Work was carried out between 1975 and 1982. Since then the plant has operated smoothly and with high availability.

An alternative to the bulb turbine is a design called the *Straflo turbine*. This is unique in that the generator is built into the rim of the turbine runner, allowing the unit it operate in low-head conditions while keeping most of the generator components out of the water. A single large Straflo turbine generator was installed at the Annapolis tidal power plant at Annapolis Royal in the Bay of Fundy, Canada. This 18 MW unit is the only one of similar size that has been built so experience with the design in limited.

Speed regulation

The speed of a conventional turbine generator has to be closely regulated so that it is synchronised with the electrical transmission system to which it is attached. In order to aid frequency regulation under the variable conditions of a tidal power plant, a set of fixed blades called a *regulator* are often placed in front of the turbine blades to impart a rotary motion to the water. The use of these blades in conjunction with a variable-blade Kaplan turbine provides a considerable measure of control over the runner speed.

In small applications where such tight speed control may not be essential and where costs are critical it may be possible to use one method of control – either a variable-blade turbine or a regulator – rather than both. An isolated unit could operate without regulation.

An alternative option is to use a variable-speed generator. This electronic solution will permit the turbine to run at its optimum speed under all conditions while delivering power at the correct frequency to the grid. This allows some efficiency gains. However the solution tends to be more costly than a conventional generator with mechanical speed control of the turbine.

Sluices and shiplocks

The sluices in a tidal barrage must be large enough and efficient enough to allow the tidal basin behind the barrage to fill with water quickly. Unless the water level behind the barrage effectively follows that on the seaward side, the efficiency of the plant is reduced.

Where the water is sufficiently deep, efficient sluices can be built using the concept of the Venturi tube. Such a design will transfer water through the barrage extremely efficiency but it must be completely submerged. More conventional sluice gates usually need to be larger than venturi tubes to provide the same rate of transfer.

Many of the rivers suitable for tidal development carry significant ship-borne trade and water traffic. To enable ships and boats to continue to use a river, shiplocks must be included in the barrage. There must also be facilities to allow fish and other forms of marine life to pass the barrage. This is particularly important if the river is one used by migratory fish such as salmon.

Modes of operation

There are several modes of operation for a tidal power plant. The simplest involves filling the tidal basin behind the plant's barrage as the tide rises, then shutting the sluice gates to prevent the water escaping. The tide is then allowed to ebb until sufficient head has developed for power generation to begin. At this point water is allowed to flow through the turbines and back to the sea.

In theory it is possible to generate power both on the ebb and the flow of the tide across a tidal barrage. La Rance, in France, was designed to operate in this way. However this method of operation has not proved the most efficient and the French plant now only operates on the ebb tide.

A more profitable strategy is to generate only when the tide is ebbing, but to use the turbines to pump water from the seaward to the landward side of the barrage close to the high tide point during the flow tide. While this involves some energy expenditure, the increased head of water behind the barrage can allow up to 10% more power to be generated than is possible without pumping. This depends, however, on the site conditions and may not always prove profitable.

Environmental considerations

Construction of a barrage across a tidal river is bound to affect the conditions on both sides of the structure. Water movement patterns will be changed, sedimentation movement will be affected and the conditions at the margins of the estuary on both the landward and seaward side of the barrage will be altered. This could have a serious effect on marine and avian life.

The major effect of the barrage will be on water levels and water movement. Water levels will be altered on both sides of the barrage and the tidal reach may change behind the barrage, although the effect will be reduced as the distance from the barrage increases. Some areas which were regularly exposed at low tide will be continuously under water after the barrage is constructed. Though the volume of water flowing down the river should remain the same, patterns of movement will be changed.

Sedimentation will be affected in complex ways. The tidal waters of an estuary frequently bear a great deal of sediment. Some is brought in from the sea, some carried downstream by the river. Changes in current speeds and patterns caused by the interpolation of a barrage will affect the amount of sediment carried by the water and the pattern of its deposition. This will, in turn, affect the ecosystems that depend on the sediment.

Other areas of concern involve animal species. The effect on fish, particularly migratory species, is significant. Fish gates can be built to permit species to cross the barrage. Many can also pass through the sluice gates. However there is a danger that fish will pass through the turbines too, being injured in the process. Various methods have been explored to discourage fish from the vicinity of the turbines, with patchy success.

Many birds live on mud flats in estuaries. There is a possibility that such mud flats would disappear after a barrage had been built, and with them the birds whose habitat they formed. Salt marshes adjacent to estuaries are also likely to be affected. Studies have been conducted on potential UK barrage sites but much work remains to be done in this area.

Against these potentially adverse effects should be balanced with the absence of any emissions such as carbon dioxide, sulphur dioxide and the oxides of nitrogen. Unlike a traditional hydropower scheme, there is little possibility of generating methane within the reservoir of a tidal plant. A tidal plant is also a sustainable source of electricity.

Global experience with tidal power plants is limited. Nevertheless, and in spite of the caveats expressed above, the evidence available suggests that such projects need have no major detrimental effect on the environment. The evidence from La Rance, in particular, has provided no serious cause for alarm. Even so it would dangerous to make any assumptions. An extremely careful environmental impact assessment would form a vital part of any future tidal project.

Financial risks

The main area of risk in tidal power development is associated with the construction of the plant and barrage. Each site is unique and will consequently present unique challenges which will need to be overcome. A careful survey will be the first requirement in order to determine the optimum site for construction of the barrage. This will need to include an accurate hydrological survey; factors such as sedimentation will depend on the precise location of the structure.

Barrage construction, particularly where a large barrage is required to dam a deep estuary, presents the greatest technical challenge. There is no consensus about how such a structure should be erected. Caisson construction appears to offer the best option although novel techniques, such as that suggested for construction of the Mersey barrage discussed above, should not be ruled out.

A tidal power plant, like a conventional hydro project, will be subject to geological risk associated with the nature of the rock foundations at the project site so an extensive geological survey will be vital. When construction begins, contingencies should be made to take into account other variables in the estuary environment including spring tides and the possibility of a river flood. Major operations such as caisson placement should be scheduled for neap tides when conditions will be at their most benign.

Compared to the uncertainty associated with barrage construction, the electromechanical equipment for tidal power should present little risk. Standard low-head turbine and generator designs are now well understood and although more advanced designs such as the Straflo turbine exist, there is no compelling reason to employ them.

One major risk associated with conventional hydropower projects, that of hydrological risk, is absent in the tidal power development. While the flow of water in a river cannot be predicted with any certainly, the tidal movement can be determined with great precision. Thus plant output can be predicted with ease.

The cost of tidal power

A tidal power plant is perhaps the most capital-intensive type of power station yet envisaged. It involves building a low-head hydropower scheme in the tidal reaches of an estuary, an environment where construction is, at best, difficult. Construction schedules are long so lengthy up-front loans are required, with a considerable gap between granting of the loan and income from the plant.

There is so little experience with this type of project that no useful conclusions can be drawn from experience. However several projects have been examined and costed, particularly in the UK and more recently in Australia. These provide some economic guidance.

The best site in England is the estuary of the River Severn. It has been extensively studied. The design favoured by the Severn Barrage Development Project in a 1989 report involved construction a power station with an installed generating capacity of 8640 MW. This was expected to take 10 years to build at a cost of around $17 billion at 1994 prices, a unit cost of $1970/kW.

A smaller project, on the River Mersey in northwest England, has also been examined in some detail. A plant with a proposed generating capacity of 700 MW was expected to cost around $1.5 billion to build, at 1994 prices, a unit cost of $2150/kW. This scheme would take 5 years to complete.

Capital costs for these two schemes are in line with the cost of similarly sized traditional hydropower projects. But tidal power has two special features which must also be taken into account. First, the load factor is low. A plant operating on the ebb tide will only generate power for half the time. Typical load factors for tidal power plants are around 23%. Efficiency can be improved slightly by pumping water from the sea across the barrage at high water to increase the head of water. However this involves additional capital expense for pump turbines.

The second special feature associated with tidal power relates to the time at which power is generated. Generation is restricted to the period between high tide and low tide. This period will occur at a different time each day.

As a consequence, the primary role of a tidal power plant is likely to be to replace generation from conventional fossil-fuelled power stations. When a tidal plant is generating, fossil fuel consumption can be cut back. When it stops generating the conventional plants must be brought back into service.

There is a way of re-timing the output of a tidal plant, but that involves building an allied electricity storage station. This would permit tidal power to be delivered either at a steady rate, or at times when the plant is not actually generating. However the addition of a storage facility pushes up the cost of the tidal project.

These features mean that the electricity generated from a tidal power station tends to be expensive. UK estimates, based on figures published by the Energy Technology Support Unit, suggest a generation cost of around $0.1/kWh assuming a discount rate for loan repayment of 8%. The cost of electricity roughly doubles if the discount rate is 15%.

In Australia, the government of Western Australia commissioned a report into a tidal power plant at Derby.[1] The report found that the most cost-effective option was a 5 MW tidal plant which would cost A$34 million. The cost of power would be A$0.41/kWh. In this case the plant was intended to replace power generated using diesel engines, which is an expensive source. However even with a renewable energy credit, the project was judged too expensive.

As both the UK and Australian examples indicate, on a purely economic basis tidal power looks uncompetitive today. But other criteria should be taken into account when determining the true cost of a tidal power plant. The lifetime of a tidal barrage is probably 120 years; and that is a conservative estimate. Turbines will probably need replacing after 30 or 40 years. Thus, once loans have been repaid, the plant will still have a long life during which it can be expected to generate cheap electric power.

Today, however, the capital intensity is crucial. And without some sort of government support or encouragement, tidal power does not look attractive. Private sector companies have been heavily involved in studies of tidal projects in the UK, but none has yet been tempted to commit itself to construction. The Derby project in Australia was also put forward by the private sector, but this was rejected by the state government in favour of fossil fuel power. While the Australian developers remain hopeful that they can develop a project elsewhere, the outlook for tidal power is generally poor. Changes in the financial and political climate may make tidal power look more attractive in the future. But for now most projects look set to remain paper studies.

End note

1 Study of Tidal Energy Technologies for Derby, prepared by Hydro Tasmania, Report No. WA-107384 – CR-01 (December 2001).

10 Storage technologies

The storage of electricity offers significant benefits for the generation, distribution and use of electric power. At the utility level, for example, a large energy storage facility can be used to store electricity generated during off-peak periods – typically overnight – and this energy can be delivered during peak periods of demand when the marginal cost of generating additional power can be several times the off-peak cost.

Energy storage plants can supply emergency back-up in case of power plant failure, helping to maintain grid stability. On a smaller scale, they can also be employed in factories or offices to take over in case of a power failure. Indeed in a critical facility where an instantaneous response to loss of power is needed, a storage technology may be the only way to ensure complete stability.

Energy storage also has an important role to play in the generation of electricity from renewable energy. Many renewable sources such as solar, wind and tidal energy are intermittent and their output often cannot be predicted with accuracy. Combining some form of energy storage with a renewable energy source helps remove this uncertainty and increases the value of the electricity generated.

Given these arguments in favour of energy storage, it may come as a surprise to learn that the use of storage plants is not widespread. One reason for the relatively small number of such plants is the availability of the technology. Another is cost. Until the late 1970s there was really only one large-scale energy storage technology and that was pumped storage hydropower. This is effective, but expensive. Since the 1980s other technologies have been developed for both utility and consumer applications but cost is still perceived as a handicap.

Yet since the 1980s there have been powerful arguments in favour of expanding storage capacity everywhere. A grid with a storage capacity of 10% to 15% of its generating capacity is much more stable and much cheaper to operate than one with virtually no storage capacity. Peaking capacity can be virtually eliminated and capacity additions can be planned more easily. But in a competitive, deregulated energy market the economics of energy storage may not appear obviously advantageous. It is probably this that has prevented greater investment.

Types of energy storage

Electricity normally has to be used as soon as it has been generated. This is why grid control and electricity dispatching systems are important; they have to balance the demand for electricity with the supply. Once one fails to match the other, problems occur. It would seem obvious, given this situation, that some reservoir of saved electricity would be a major boon to grid operation. Yet storing electricity has proved difficult to master.

Storing electricity in its dynamic form, amperes and volts, is almost impossible. The nearest one can get is a superconducting magnetic energy storage ring which will store a circulating DC current indefinitely provided it is kept cold. A capacitor storage system stores electricity in the form of electric charge. All other types of energy storage convert the electricity into some other form of energy. This means that the energy must then be converted back into electricity when it is needed.

A rechargeable battery may appear to store electricity but in fact it stores the energy in chemical form. A pumped storage hydropower plant stores potential energy; a flywheel stores kinetic energy while a compressed air energy storage (CAES) plant stores energy in the form of compressed air, another type of potential energy. Alternatively one might use electrolysis to turn electricity into hydrogen, yet another chemical form of energy.

All these, and one or two others, represent viable ways of storing electricity. Several are commercially available, others in the development stage. And each has its advantages and disadvantages.

For large-scale utility energy storage there are three possible technologies to chose between, pumped storage hydropower, CAES and, at the low end of the capacity range, large batteries. Batteries can also be used for small- to medium-sized distributed energy storage facilities,[1] along with flywheels and capacitor storage systems. Superconducting magnetic energy storage is being used for small storage facilities and would be suitable for large facilities but is prodigiously expensive.

Some of these systems can deliver power extremely rapidly. A capacitor can provide power almost instantaneously, as can a superconducting energy storage system. Flywheels are very fast too, and batteries should respond in tens of milliseconds. A CAES plant probably takes 2–3 min to provide full power. Response times of pumped storage hydropower plants can vary between around 10 s and 15 min.

The length of time the energy must be stored will also affect the technology choice. For very long-term storage of days or weeks, a mechanical storage system is the best and pumped storage hydropower is the most effective provided water loss is managed carefully. For daily cycling of energy, both pumped storage and CAES are suitable while batteries can be used to store energy for periods of hours. Capacitors, flywheels and superconducting magnetic energy storage are generally suited to short-term energy storage, though flywheels can be used for more extended energy storage too.

Table 10.1 *Round trip energy efficiencies for storage technologies*

	Efficiency (%)
Capacitors	90
Superconducting energy storage	90
Flow batteries	90
CAES	80
Flywheels	80
Pumped storage hydropower	75–80
Batteries	75–90

Another important consideration is the efficiency of the energy conversion process. An energy storage system utilises two complementary processes, storing the electricity and then retrieving it. Each will involve some loss. The round trip efficiency is the percentage of the electricity sent for storage which actually reappears as electricity again. Typical figures for different types of system are shown in Table 10.1.

Electronic storage systems such as capacitors can be very efficient, as can batteries. However the efficiencies of both will fall with time due to energy leakage. Flow batteries, where the chemical reactants are separated, perform better in this respect and will maintain their round trip efficiency better over time. Mechanical storage systems such as flywheels, CAES and pumped storage hydropower are relatively less efficient. However the latter two, in particular, can store their energy for long periods if necessary without significant loss.

Pumped storage hydropower

The most widespread large-scale electricity storage technology is pumped storage hydropower. This is also the oldest storage technology in use, with the first plant built at the beginning of the twentieth century. By the beginning of the twenty-first century there was probably 140,000 MW of pumped storage capacity in operation.

A pumped storage plant is like a conventional hydropower plant with a dam and reservoir but in this case there are two reservoirs rather than one. These two reservoirs must be separated vertically; one must be higher than the other. The difference in height provides the head of water to drive the station's turbines.

In order to generate power, water runs from the top reservoir through a high-pressure channel to turbines at the bottom of the drop. The turbines extract the potential energy from the water and then discharge it into the bottom reservoir where it is saved. When energy is to be stored, the turbines are reversed and act as pumps, pumping water from the lower reservoir

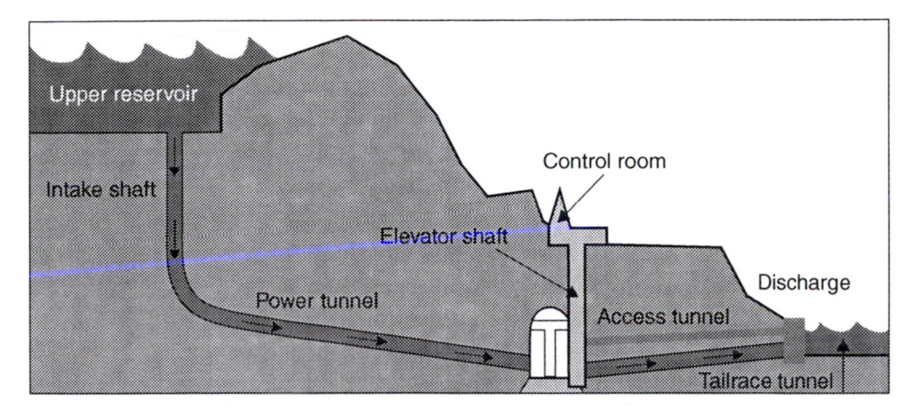

Figure 10.1 *Cross section of a pumped storage hydropower plant*

into the upper. The turbine pumps are driven using off-peak electricity so storage will normally take place at night. Once water has been pumped into the upper reservoir it is available again for power generation.

This type of plant is extremely robust and though round trip efficiency is lower than for some other technologies, long-term energy losses are low. Leakage and evaporation are the main sources of loss and if these are managed well, water loss can be kept small.

Plant design

The energy that can be extracted from a hydropower plant depends on both the volume of water available and the head of water that can be exploited. A pumped storage project will provide the most efficient and cheapest operation when it can provide a high head between its two reservoirs. This will allow the greatest amount of energy to be stored in the smallest volume of water. That, in turn, means that pumps and turbines can be smaller, reducing the capital cost of the plant.

Turbines

The earliest pumped storage power plants used separate turbines and pumps but this made the projects costly to build. The development of reversible pump turbines made the economics of the pumped storage plant look much more attractive.

Most reversible pump turbines used in storage plants are Francis turbines. The Francis design is well suited to both generation and pumping and can pump water to a considerable height (see Chapter 8 for more detail about hydropower turbines).

There is one drawback with the Francis design for this application; the turbine blade angle is fixed. A fixed blade does not provide the best efficiency for both pumping and generating power. An alternative design called the Deriaz turbine, similar to the Francis turbine in design but with movable blades, has been used in several pumped storage projects to try and achieve greater efficiency.

Propeller-like Kaplan turbines (see Chapter 8) can also be used as pumps, though not to transfer water to a reservoir of any great height. The La Rance tidal power plant in France, for example, uses such turbines to pump water across its tidal barrage in order to increase efficiency of operation of this station.

The best efficiency that a hydraulic turbine can provide for generating power is around 95%. Pumps are less efficient, operating at best at around 90%. This means that the best efficiency that can be expected from a pumped storage power plant through a storage and regeneration cycle is around 86%. In practice the efficiency is normally between 75% and 80% as shown in Table 10.1.

Francis and Deriaz turbines can be built today to operate at heads of up to 700 m in a single stage. Beyond that it will usually be necessary to use a combination of a pump and a Pelton turbine. Several plants in Switzerland employ this configuration.

Pumped storage hydropower plants can be brought on-line extremely quickly. The 1800-MW power station at Dinorwig in Wales, for example, can be run up from zero output to 1800-MW output in around 10 s. This ability makes pumped storage extremely attractive as a system reserve to be brought into service if a major base-load unit breaks down.

Global exploitation

Of all the energy storage technologies in use, pumped storage hydropower is by far the most widely adopted. Since its introduction in Switzerland in around 1904, plants have been built in other parts of Europe, in the USA, in China, in Japan and in many other countries.

Pumped storage facilities have often been built in conjunction with nuclear power plants. This combination allows the nuclear plant to run continuously at full power, its most effective mode of operation. Electricity from the generating plant not required immediately by the grid is stored for dispatching during peak demand periods. Nuclear power plants have generating capacities up to 1300 MW; only a pumped storage plant can provide the large storage capacity needed in this situation.

The two countries with the largest pumped storage hydropower capacities are Japan and the USA. Each has around 20,000 MW. In both cases many of these plants are associated with nuclear development. In the USA, for example, the bulk of the capacity was built between 1970 and 1990 when nuclear growth was greatest. There is a further 100,000 MW distributed across the globe, providing a global capacity or close to 140,000 MW.[2]

Financial risks

Pumped storage employs the same technology and construction techniques used in conventional hydropower projects and the risks are similar. These fall into three groups: geological, hydrological and technical risks.

Geological risk will depend on the site for the project. This will have to be capable of providing two reservoirs and room for a power station. In some cases the sea can be used as the lower reservoir, simplifying the design. As with all hydropower schemes, a thorough feasibility study is vital to assess the geological conditions. Faults within the underlying rock structure could cause construction problems, leading to major cost overruns if not identified early. Risk of seismic shock must also be considered.

The hydrological risk associated with a pumped storage hydropower plant should be slight since the station will not normally depend on a supply of water from a river which may be unpredictable. However any problems with water loss from evaporation or through leakage from the reservoirs will affect plant economics. Technical risk is minimal too. Hydro turbine technology is well established and should not lead to any problems.

Costs

Capital costs are likely to be broadly in line with those for a conventional hydropower project – a unit cost of between $800/kW and $3,500/kW – although the specialised nature of the pump turbines, or the need for separate pumps and turbines could push the cost of the plant up. Some pumped storage plants place the lower reservoir underground. This is likely to increase construction costs.

Compressed air energy storage

CAES is exactly as its name suggests; air is compressed and stored under pressure. Release of the pressurised air is subsequently exploited to generate electricity. Although the storage of compressed air is clearly a means of storing energy, it is only when it is considered in conjunction with the gas turbine that it makes complete sense from a power generation perspective.

A gas turbine consists of two major components. These are a compressor and a turbine. Conventional gas turbines used in aero applications or for power generation have the two components mounted on a single drive shaft.

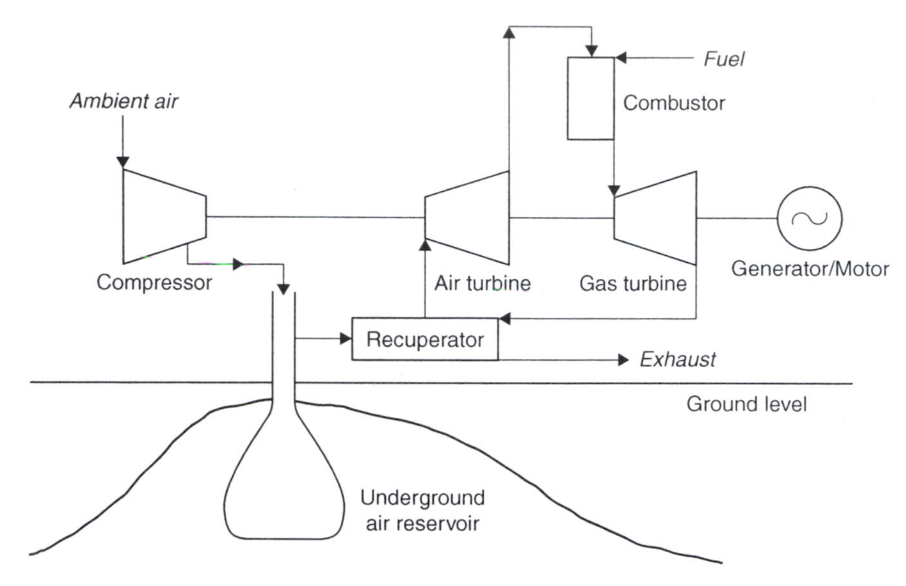

Figure 10.2 *Diagram of a CAES plant*

During conventional operation, air is drawn into the compressor and compressed. This compressed air is then directed into a combustion chamber where it is mixed with fuel and ignited. Heating the compressed air increases its energy content significantly. The hot compressed gas is then released through the machines turbine blades, causing them to rotate and generate electricity or motive power.

Although a gas turbine normally has the compressor and turbine closely integrated, there is no reason, in principle, why compression should not be carried out separately, and at a different time to power generation. This is the crux of the CAES plant.

In a CAES plant the compressor and the turbine are separated. By use of a system of clutches, each can be linked, separately, to a motor generator. In storage mode the compressor stage of the gas turbine is driven by the reversible motor generator using off-peak power from the grid system. The product, compressed air, is stored in a special cavern.

When the power is required, air is released from the cavern into a combustion chamber, mixed with fuel, ignited and allowed to expand through the turbine section of the system. Under these conditions the motor generator is used in generation mode to produce electricity.

Storage caverns

The most important part of a CAES plant is somewhere to store the compressed air. Small-scale CAES plants – with storage capacities of up to

20 MWh – can use overground storage tanks but large, utility-scale plants need underground caverns in which to store the air. The natural gas industry has used underground storage caverns for years to store gas; these same caverns can provide ideal storage facilities for a CAES plant. However the demand for such a cavern limits the development of CAES to places where such storage caverns are available.

A number of different types of underground cavern can be exploited. The simplest is a man-made rock cavern. This must be sited in an impervious-rock formation if it is to retain the compressed air without loss.

Salt caverns have been commonly used for gas storage. These are created by dissolving or dry mining salt to create a suitable enclosure. Salt deposits suitable for such caverns occur in many parts of the world.

A third type of underground storage is found in porous rock bounded by an impervious barrier. Examples can be found in water-bearing aquifers, or as a result of oil and gas extraction. Aquifers can be particularly attractive as storage media because the compressed air will displace water, setting up a constant pressure storage system. With rock and salt caverns, in contrast, the pressure of the air will vary as more is added or released.

All three types of storage structure require sound-rock formations to prevent the air from escaping. They also need to be sufficiently deep and strong to withstand the pressures imposed on them. It is important, particularly in porous-rock storage systems, that there are no minerals present that can deplete the oxygen in the air by reacting with it. Otherwise the ability of the air to react with the fuel during combustion will be affected, reducing the power available during the generation phase of the storage generation cycle.

Turbine technology

A CAES plant uses standard gas turbine compressor and turbine technology but because the two units operate independently, they can be sized differently in order to match the requirements of the plant. The larger the compressor compared to the turbine, the lesser the time it requires to charge the cavern with a given amount of energy. Thus a plant built in Germany required 4 h of compression to provide an hour of power generation whereas a plant in Alabama needs only 1.7 h of compression for an hour of generation.

As a result of compression and generation being separated, a CAES plant turbine can operate well at part load as well as full load. More complex operation is also possible. The Alabama plant, for example, uses two turbine stages with the exhaust from the last turbine used to heat air from the cavern before it enters the first turbine.[3] Fuel is not actually burnt in the compressed air until it enters a combustion chamber between the first and second turbine stages.

A key feature of a CAES plant is that it generates more electricity than was actually consumed when the air was stored. This is a result of the fuel burnt in the compressed air during the generation part of the cycle. A typical plant will deliver 30–35% more electricity to the grid that was originally consumed during storage.

Global exploitation

CAES has had a short history of limited development. The largest project yet built was a 290-MW power plant constructed at Huntorf in Germany in 1978. This plant operated for 10 years with 90% availability and 99% reliability, providing storage for a nuclear plant. Even so the German utility decommissioned the project. Interest in CAES then shifted to the USA where the Electric Power Research Institute (EPRI) began to promote the technology in the latter half of the 1980s.

EPRI saw CAES as a useful technology to enable small US utilities limit their need for expensive peaking power stations. It estimated that rock formations capable of providing reliable underground storage caverns exist across 75% of the USA .

As a result of EPRI's work, a 110-MW commercial project was built by the Alabama Electric Cooperative. The plant entered service in May 1991 and has operated ever since. It cycles once or twice each day, and can store 2600 MWh of energy.

At around the time the Alabama plant was built, Italy tested the technology in a 25-MW installation. No plant was built. More recently a 2700-MW plant has been proposed in the USA but not yet constructed.

Financial risk

There should be little technical risk associated with a CAES project. Gas turbine technology is well understood and relatively cheap. Gas storage techniques are also well tested. The combination of the two remains novel but practical experience suggests that the technique is both robust and reliable.

Costs

There is little experience with CAES so any cost estimate must be considered tentative. However it would appear to be an economically attractive option for energy storage. Installation costs of around $400/kW have been mooted for the USA.

Large-scale batteries

The traditional way of providing electricity storage has been the battery. This is an electrochemical device which stores energy in a chemical form so that it can be released as required.

A battery comprises a series of individual cells, each of which is capable of providing a defined current at a fixed voltage. Cells are joined both in series in and parallel to provide the required voltage and current rating required for a particular application.

Each cell contains two electrodes, an anode and a cathode. These are immersed in an electrolyte.[4] At its simplest, the electrodes are made of materials which will react together spontaneously but the electrolyte in which they are immersed will allow the passage of only one of the components required to complete the reaction.[5] An electrical connection must be made between the two electrodes to allow the passage of electrons from one electrode to the other in order to complete the reaction. This is the source of electrical power.

There are two different types of traditional batteries: the primary cell and the secondary cell. A primary cell can only be discharged once, after which it must be discarded. A secondary cell can be discharged and recharged many times. Only the second type is of any use for energy storage systems.

Secondary cells can further be divided into shallow discharge and deep discharge cells. A shallow discharge cell is only partially discharged before being recharged again; an automotive battery would typify this type of cell. A deep discharge cell is normally completely discharged before recharge. This is the type which is most attractive for large-scale electricity storage.

Traditional electrochemical storage systems boast a best case conversion efficiency of 90% but a more typical figure would be 70%. Most batteries also suffer from leakage of power. Left for too long, the cell discharges itself. This means that battery systems can only be used for relatively short-term storage.

An additional problem with batteries is their tendency to age. After a certain number of cycles, the cell stops holding its charge effectively, or the amount of charge it can hold declines. Much development work has been aimed at extending the lifetime of electrochemical cells but this remains a problem.

To their advantage, batteries can respond to a demand for power almost instantaneously. This property can be used to good effect to improve the stability of an electricity network. It is also valuable in both distributed generation and for back-up power applications.

Traditional batteries are completely self-contained. However there is another type called a *flow battery* in which chemical reagents involved in the generation of electricity are held is tanks separated from the actual electrochemical cell. In this type of device the reagent is pumped through the cell as needed. Such cells suffer less from energy leakage. Several types are being developed for utility electricity storage.

Lead acid batteries

Lead acid batteries are the best known of all rechargeable batteries. These are the cells used in automobiles worldwide as well as for small-scale energy storage in homes and offices. Advanced lead acid cells have been developed for utility storage applications, the largest being for a 10-MW plant in California.

Lead acid batteries operate at ambient temperature and use a liquid electrolyte. They are extremely heavy and have a poor energy density but neither of these is a major handicap for stationary applications. They are also cheap and can be recycled many times, though they should not be completely discharged as this can cause problems.

Nickel–cadmium batteries

Nickel–cadmium batteries have higher-energy densities and are lighter than lead acid batteries. They also operate better at low temperatures. However they tend to be more expensive. This type of battery was used widely in portable computers and phones but has now been superseded by lithium ion batteries. The largest nickel–cadmium battery ever built is a 40-MW unit in Alaska which was completed in 2003. It occupies a building the size of a football field and comprises 13,760 individual cells.

Sodium–sulphur batteries

The sodium–sulphur battery is a high-temperature battery. It operates at 300°C and contains liquid sodium which will explode if allowed in contact with water. Safety is a major issue with these batteries. However the battery has a very-high-energy density which makes it attractive, particularly for automotive applications.

The battery is being developed for utility applications in Japan. Demonstration and early commercial projects have ranged in size from 500 kW to 6 MW. Most are in Japan but a small unit was commissioned in the USA in 2002.

Flow batteries

A flowing-electrolyte battery, or flow battery is a cross between a conventional battery and a fuel cell. It has electrodes like a conventional battery where the electrochemical reaction responsible for electricity generation or storage takes place and an electrolyte. However the chemical reactants responsible for the electrochemical reaction and the product of that reaction

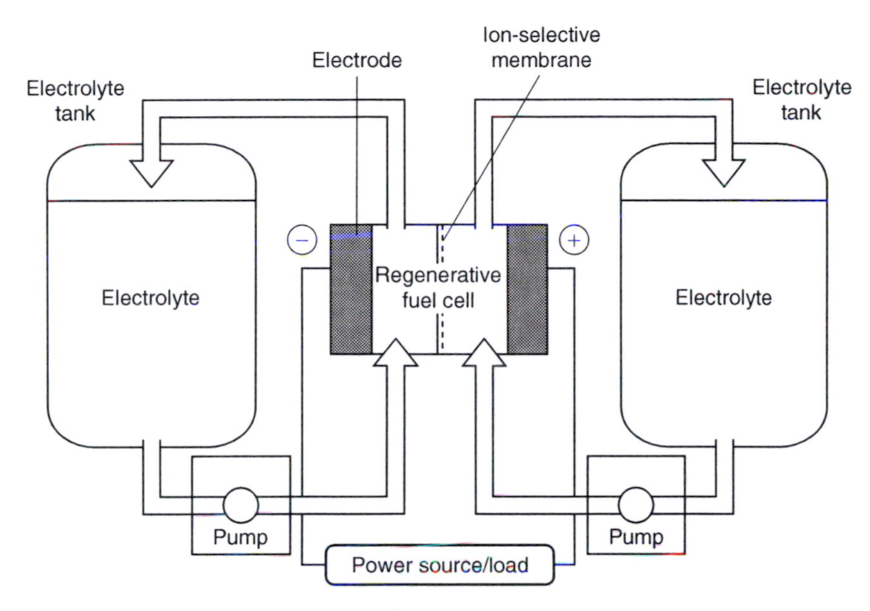

Figure 10.3 *Diagram of a typical flow battery*

are stored in tanks separate from the cell and pumped to and from the electrodes as required, much like a fuel cell.[6]

Two types of flow battery have been developed for utility applications, the polysulphide-bromide battery and the vanadium redox battery. Both designs have been developed to the demonstration stage. Capacities of up to 15 MW have been proposed. Response time from zero to full power is expected to be around 100 ms.

Financial risks

While battery technology is over a century old, the types of cell proposed for utility storage are novel and experience with them is limited. Most of the promising designs are in the demonstration of early commercialisation stage. This uncertainty about the technology means that there is a significant technological risk. Some operating lead acid storage plants are now over a decade old, providing early feedback about cell lifetime as well as operating experience. Much more is needed to establish a good measure of their potential.

Costs

Initial estimates suggest that lead acid batteries will cost around \$500/kW to install. Sodium–sulphur batteries are expected to cost around \$1000/kW while flow batteries should cost between \$800/kW and \$900/kW. Costs for both can be expected to fall if demonstrations prove successful and lead to commercial uptake.

Superconducting magnetic energy storage

Superconductivity offers, in principle, the ideal way of storing electric power. The storage system comprises an electromagnetic coil of superconducting material which is kept extremely cold. Off-peak electricity is converted to DC and fed into the storage ring, and there it stays, ready to be retrieved as required. Provided the system is kept below a certain temperature, electricity stored in the ring will remain there indefinitely without loss.

The key to the superconducting magnetic energy storage device is a class of materials called *superconductors*. Superconductors undergo a fundamental change in their physical properties below a certain temperature called *the transition temperature* which is a characteristic of each material. When a material is cooled below its transition temperature it becomes superconducting. In this state it has zero electrical resistance. This means that it will conduct a current with zero energy loss.

Unfortunately the best superconducting materials only undergo this transition at below 20°K (−253°C). Temperatures this low can only be maintained by cooling the superconducting coil with liquid hydrogen or liquid helium, in either case an expensive process.

In recent years scientists have discovered materials that become superconducting at relatively high temperatures, temperatures accessible by cooling with liquid nitrogen. (Liquid nitrogen boils at 98°K, −175°C.) Most of these materials have proved to be rather brittle ceramics which are difficult to work but techniques are being found to exploit them. This is helping make superconductivity more economically attractive for a range of utility applications including storage.

Superconductors store DC current without loss but losses occur in converting the off-peak AC current to DC and then back to AC when required. The round trip efficiency is around 90%. A superconducting magnetic storage device can respond extremely quickly, delivering its rated power in about 20 ms.

A number of small superconducting storage rings have been built for use as power-conditioning systems. One of 10 MW capacity has been tested on a utility system in the USA where its primary role was to improve transmission system stability. Such systems are extremely expensive.

The unit cost of storing power in a superconducting ring decreases as the size of the plant increases so large storage devices would be preferred for utility applications. The superconducting ring for a 5000-MW device would be roughly 1600 m in diameter. The magnetic fields associated with such a device would be enormous and it would have to be built into rock to ensure it did not collapse under the force generated.

Financial risks

Superconducting magnetic energy storage involves a range of advanced technologies, most of which have not been proved beyond the experimental or small-scale demonstration stage. New, high-temperature superconducting materials are being developed but these remain experimental. The technological risk associated with the use of this technology is high. It is unlikely to be appropriate for wide scale use for several years.

Costs

Superconducting rings are costly to make and costly to operate. Although figures of around \$2000/kW have been mooted for 1000 MW installations, smaller units cost in excess of \$3000/kW.

Flywheels

A flywheel is a simple mechanical energy storage device comprising a large wheel on an axle fitted with frictionless bearings. A flywheel stores kinetic energy as a result of its rotation. The faster it rotates, the more energy it stores. In order for a flywheel to be effective as an energy storage device there must be a way of feeding energy into the flywheel and a means of extracting it again.

Simple flywheel energy storage devices are fitted to all piston engines to maintain smooth engine motion. The engine flywheel is attached physically to the engine camshaft and as the pistons cause the camshaft to rotate they feed energy into the flywheel. For electricity storage applications, energy will normally be fed into the flywheel using a reversible motor generator.

The faster a flywheel rotates, the more energy it will store. Conventional flywheels are fabricated from heavy metal discs made of iron or steel. However these discs are only capable of rotating at low speeds. For power applications, new lighter composite materials are being developed, capable of rotating at 10,000–100,000 rpm without fracturing under the immense centrifugal force they experience. Such devices must be housed in exceedingly strong containers which will prevent the pieces of the flywheel scattering like shrapnel in the event of a catastrophic failure.

Energy storage systems must operate with low energy loss. This is accomplished in flywheel systems by using magnetic bearings to eliminate bearing friction and by operating the flywheel in either a vacuum or in a container filled with a low-friction gas such as helium.

One of the problems with flywheel energy systems is that the flywheel will rotate at varying speeds depending on how much energy it contains. If a conventional motor generator is used to extract electrical energy from the flywheel, this will translate into a variable frequency output. Grid electricity, however, must be generated at a constant AC frequency. Various electromechanical and electronic means of overcoming this difficulty have been found.

Flywheels have the attraction of virtually zero maintenance and infinite recyclability. They have proved to be one of the best and cheapest ways of maintaining power quality during power failure or network voltage or frequency dip. Response time is fast and in the case of power failure a flywheel system can bridge the period between the power outage and a long-term back-up system such as a generator set coming on line.

The largest flywheel system so far built is a 1-MW unit comprising 10 100-kW flywheels used to maintain system voltage on the New York transit system. Storage capacity is 250 kWh, sufficient to provide 1 MW for 15 min.

Financial risks

Flywheels represent a conventional technology extended by the use of unconventional materials. They have been under development for many years. Units are now available commercially and appear to be predictable. The technological risks associated with this type of storage technology would appear to be fairly low.

Costs

The capital cost of a flywheel storage system may be as high as $2000/kW, though costs should fall below this for standard modular units. However against this high cost must be balanced the fact that they are virtually maintenance-free, can be cycled indefinitely and are extremely predictable. The energy contained in the system can always be determined. In applications where medium capacity, short-term storage is required, flywheels offer one of the best solutions.

Capacitors

Capacitors are used extensively in electrical and electronic circuitry. In power networks they have been used to enhance system stability. More advanced capacitors are now being developed specifically for energy storage.

The classic capacitor comprises two parallel metal plates with an air gap between them. When a voltage is applied to the plate a positive charge collects on one plate and a negative charge on the other.

A number of different capacitor types exist. Those being considered for energy storage are called *electrochemical capacitors*. These utilise a solid electrode and an electrolyte. Charge collects at the interface between the two. These devices, sometimes called *super capacitors* or *ultra capacitors* can store a very large energy density, probably the highest of any storage device. They can respond in tens to hundreds of milliseconds and are most suited to short-term energy storage applications.

The technology is relatively new and there is little cost data available. Nor is lifetime or operational experience available, but static electrical devices of this type should show good long-term stability and should be relatively maintenance-free.

Hydrogen

When a sufficiently high voltage is applied to water using two electrodes, the effect is to cause the water to decompose into its two elemental constituents, hydrogen and oxygen. One gas appears at one electrode, the other at the second, so it is relatively simple to separate the two. This can form the basis of an energy storage system.

When water is electrolysed in this way the oxygen is normally discarded while the hydrogen is retained. Hydrogen is an excellent and versatile fuel which can be burnt cleanly in a power plant to regenerate electricity or used in a variety of other ways such as fuel for motor vehicles. Ideally the hydrogen would be burned in a fuel cell, a device capable of up to 60% energy conversion efficiency – perhaps rising 75% in a combined cycle configuration. When burnt, the product of combustion is water.

A major problem with hydrogen as a storage medium is round trip efficiency. Hydrolysis of water is generally only around 70% efficient, though some companies have claimed up to 80%. Assuming hydrogen to electricity conversion efficiency of 75%, and ignoring other losses, the round trip efficiency would be 60%. This is optimistic; when storage and other losses are taken into account, it would probably result in a round trip efficiency of closer to 50%.

While this is obviously a major handicap, the advantages of hydrogen may eventually make such losses acceptable. Fuel technology based on hydrogen is being developed but is not currently commercially viable. However the concept of a hydrogen economy[7] could take off later this century.

Environmental considerations

Each of the energy storage technologies considered in this chapter has an environmental impact related to the technology and techniques it employs. Pumped storage hydropower, for example, will entail many of the same considerations that apply to conventional hydropower while CAES will involve similar emission considerations as those relevant when considering a gas turbine.

Large battery energy storage systems may involve the use of toxic materials such as cadmium or lead which need to be handled and eventually recycled with care. The sodium in a sodium–sulphur battery is particularly dangerous if not handled carefully. Flow battery systems contain reagents which should not be allowed to escape into the environment either.

High-technology storage systems such as superconducting magnetic energy storage and super capacitors will also involve novel, possibly toxic materials. However these will usually be costly to produce and there will be a strong incentive to recycle them. Flywheels are probably the most benign of the technologies with little environmental impact unless treated extremely carelessly.

There are, however, two aspects of storage technologies that have wider ranging impacts. The first is their ability to improve overall system efficiency and the second is the advantages that accrue to their use in conjunction with renewable technologies.

Adding energy storage capacity to a transmission or distribution network makes it easier to manage. As already indicated, storage capacity can be used to store off-peak electricity generated in cheap base-load generating plants, electricity that can then be used when demand rises beyond the capacity of the base-load units.

This mode of operation is economical because it replaces peak-load generation with base-load generation and the latter is normally the cheapest source of electricity. It is also more efficient since it will allow a utility to base the majority of its generation on its most energy efficient units. This is of environmental benefit since the most efficient generation results in the lowest atmospheric pollution – or should if regulation is operating correctly.

Renewable energy

Improved energy efficiency is one consequence of the use of energy storage. However electricity storage can also have a profound effect on the economics and utility of renewable energy sources. Wind power, solar power, tidal power, wave power; these are all either intermittent or unpredictable sources of electricity, or both. Both features are a handicap which makes the energy less valuable to a power network operator and less easy to accommodate in large quantities. There is a limit to the amount of unpredictable power a network can accept while still providing a good service.

If energy storage is added alongside these renewable sources, the situation becomes completely different. Now the energy from the wind or solar plant can either be used if required or stored. The output from these plants is averaged out. Both peaks and troughs are accommodated by the storage unit. As a consequence the energy source becomes predictable. This makes it much easier to dispatch and it also allows larger quantities to be accepted without affecting the quality of the network supply.

There is a price to pay and that is the cost of the storage system. At the beginning of the twenty-first century that makes the combination an uneconomical prospect. But as the cost of renewable energy drops and that of fossil fuel rises, and as the general benefits of larger energy storage capacities are accepted, the economics are likely to look less of a disadvantage.

Costs

The costs of energy storage systems vary widely. Some, like pumped storage hydropower are inherently expensive to build, while others, like superconducting magnetic energy storage are expensive because they are new. One or two, like CAES, are relatively cheap.

Table 10.2 presents some tentative costs for the different technologies discussed here. They suggest that as already noted, CAES is the cheapest to install though battery storage could also be inexpensive. The figures should be treated with caution, however, particularly because many of the technologies are under development and costs are likely to fall significantly once they become widely available commercially.

When considering the economics of a storage system the round trip efficiency will also be a consideration. This will determine how much of the electricity used to charge the storage plant can actually be returned to the system.

With the exception of CAES, a storage plant does not use any fuel. Thus there are usually no fuel costs to consider. Many of the technologies are relatively easy to operate and maintain too.

Overall, however it is the conversion of off-peak electricity into peak-period electricity that dominates the economics of a storage plant. It is this equation, therefore, that will determine whether the plant is economical or not.

Table 10.2 *Capital costs for energy storage systems*

	Cost ($/kW)
Flywheel	2000
Superconducting magnetic energy storage	2000–3000
Pumped storage hydropower	800–3500
CAES	400
Battery storage	500–1000

End notes

1 Distributed storage facilities may be used by utilities to improve local grid stability or they may be used by consumers to make their own supplies more secure.
2 The Commercial World of Energy Storage: A Review of Operating Facilities (under construction or planned), Septimus van der Linden, 1st Annual Conference of the Energy Storage Council, Houston, 2003.
3 This process is called recuperation and it reduces fuel consumption by around 25%.
4 This is a simplification. Some advanced cells utilise pastes or solid electrolytes.
5 Normally the electrolyte will permit a charged molecule, an ion, to pass but will not conduct electrons.
6 The processes occuring here is somewhat different to the simple electrochemical process described above but the principle is similar.
7 A hydrogen economy is one in which hydrogen replaces oil, gas and coal as the main fuel in automotive, heating and many power generation applications. It is seen as one option for a sustainable energy future.

11 Wind power

Wind is the movement of air in response to pressure differences within the atmosphere. Pressure differences exert a force which cause air masses to move from a region of high pressure to one of low pressure. That movement is wind. Such pressure differences are caused primarily by differential heating effects of the sun on the surface of the earth. Thus wind energy can be considered to be a form of solar energy.

Annually, over the earth's land masses, around 1.7 million TWh of energy is generated in the form of wind. Over the globe as a whole the figure is much higher. Even so, only a small fraction of the wind energy can be harnessed to generate useful energy.

One of the main limiting factors in the exploitation of wind power onshore is competing land use. Taking this into account, a 1991 estimate[1] put the realisable global wind power potential at 53,000 TWh/year. This figure is broken down by regions in Table 11.1. As the table shows, wind resources are widely dispersed and available in most parts of the world.

The figures in Table 11.1 are probably conservative because modern wind turbines are more efficient than those available when the survey was compiled. Even on this conservative estimate the resource is much larger than world demand for electricity. This is expected to reach 26,000 TWh, roughly half the global wind resource quoted above, by 2020.[2]

Table 11.1 *Regional wind resources*

	Available resource (TWh/year)
Western Europe	4800
North America	14,000
Australia	3000
Africa	10,600
Latin America	5400
Eastern Europe and Former Soviet Union	10,600
Asia	4600
Total	53,000

Table 11.2 *European wind energy resources*

	Annual resource (TWh)	*Potential capacity (MW)*
Austria	3	1500
Belgium	5	2500
Denmark	10	4500
Finland	7	3500
France	85	42,500
Germany	24	12,000
Great Britain	114	57,000
Greece	44	22,000
Ireland	44	22,000
Italy	69	34,500
Luxembourg	–	–
Holland	7	3500
Norway	76	38,000
Portugal	15	7500
Spain	86	43,000
Sweden	41	20,500

Source: The figures in this table are taken from Windforce 12.[3]

Table 11.2 shows estimates for the wind energy resources in the countries of Western Europe. A glance at these estimates will show that in many cases the national wind resources are again enormous. These figures, too, may represent an underestimate. For example, the potential UK generating capacity has been estimated by the UK Energy Technology Support Unit (ETSU) to be 223,000 MW, nearly four times the figure quoted in Table 11.2 and equivalent to an annual production of 660,000 GWh. The ETSU study used relatively conservative criteria to arrive at its estimate but did not take into account utilisation restrictions. Constraints on building close to population centres or in areas of natural beauty would severely limit available sites. Even with such constraints, the potential would remain vast.

Looking beyond Europe, a US wind potential survey was carried out in 1992.[4] It concluded that even with exclusions for environmental and land-use reasons, around 6% of the land area of the USA could be used for wind power generation. This area was judged capable, with some advance in wind turbine technology, of providing a generating capacity of 500,000 MW. The report also concluded that 12 states in the middle of the USA had sufficient potential to generate nearly four times the electricity consumed in the USA 1990.

In Asia, potential Chinese generating capacity has been put at 253,000 MW. Indian potential has been estimated at 20,000 MW but this is certainly a severe underestimate. Both countries are beginning to exploit their potential.

Wind power, though exploiting a renewable resource, is not considered beneficial by all. In the UK, and increasingly in Germany there are lobbies trying to prevent further development of onshore wind farms. This is proving a considerable handicap to wind development in the UK, at least. Under these conditions, offshore wind farming becomes increasingly attractive.

It is more expensive to build a wind farm offshore but this can be offset by higher average wind speeds. The global offshore resource has been estimated to be around 37,000 TWh.[5] Offshore sites are available in many parts of the world but the most promising for immediate development are around the coasts of northern and western Europe and of the eastern seaboard of the USA.

At the beginning of 2004 the global wind generating capacity was 40,000 MW.[6] It is expected to reach 150,000 by 2012. Offshore capacity at the end of 2003 was just over 500 MW.

Wind sites

The economics of wind power depend strongly on wind speed. The actual energy contained in the wind varies with the third power of the wind speed. Double the wind speed, and the energy it carries increases eightfold.

A 1.5 MW wind turbine at a site with a wind speed of 5.5 m/s will generate around 1000 MWh/year. At a wind speed of 8.5 m/s the output rises to 4500 MWh and at 10.5 m/s the annual output will be 8000 MWh. This is close the theoretical limit. Other factors will come into play at very high speeds, limiting turbine output. However these figures indicate quite clearly that the selection of a good wind farm site is vitally important for the economics of a project.

The starting point for any wind development, then, must be a windy site. But other factors come into play too. Wind speed varies with height; the higher a turbine is raised above the ground, the better the wind regime it will find. This will benefit larger wind turbines which are placed on higher towers, but larger turbines tend to be more efficient anyway, so additional advantages accrue.

Depending on the efficiency of a wind turbine, there is a cut-off wind speed below which wind power generation is not considered economical. This figure depends on the efficiency of wind turbine design as well as on the turbine cost. With the turbines available at the beginning of the twenty-first century, a wind speed as low as 5–5.5 m/s is considered economically exploitable at an onshore site. Since offshore costs are higher, an offshore wind speed of 6.5 m/s is needed to make a site economically attractive.

Locating a site

Prospective developers of wind energy projects will normally be able to refer to wind surveys in most of the developed countries in order to make a preliminary identification of sites suitable for wind farms. Wind energy associations exist in the UK, Europe and the USA and the European Union (EU) also holds Europe-wide figures. In other parts of the world the wind data that is available may be less precise, though many countries are now taking greater interest in wind resources.

Once a potential site has been identified it must be studied in more detail to confirm that it is suitable. Long- and short-term wind speed measurements will normally be needed to ascertain the wind regime. Only when these figures are available can the economics of the project be determined with any accuracy. Figures for at least one full year will normally be required, longer if possible.

Offshore projects require the same attention as onshore schemes but offshore wind data is less likely to be available. European offshore surveys exist and there have been some limited surveys of offshore North American sites. It is possible to gain an estimate of the wind regime in an offshore area from satellite images. These can provide an indication of sea roughness from which wind speed can be calculated. As with an onshore site, accurate measurements over at least a year will then be needed to confirm the local wind regime.

Turbulence

When wind passes over land the unevenness of the ground and interference to wind flow from trees or undergrowth will cause a significant amount of turbulence. Turbulent air creates an additional strain on a wind turbine blade, accelerating the onset of fatigue damage. In order to limit this damage as much as possible wind turbine designers will normally place the turbine on a tower, which is tall enough to raise the blades above this turbulent layer of air.

The wind offshore is generally less disturbed because the surface of the sea is smoother, resulting in a thinner turbulent layer and less overall turbulence. Waves in rough seas will increase turbulence and wave height itself needs to be taken into account offshore. Turbine blades should be lifted high enough to avoid the highest waves likely at a particular site. Generally, however, tower heights offshore can be lower than onshore. In both cases site measurements will be needed to ascertain what the optimum turbine height should be.

Wind turbines

Like hydropower, wind power has a long and successful history. The earliest known record is from Hero of Alexandria who described a wind machine in the first century AD. The next recorded appearance is in Arabic texts of the ninth century which refer to a seventh century design.

Windmills soon spread from the Middle East into Europe. Post mills, in which the whole mill apparatus was mounted on a post so that it could be rotated into the wind, were known in France and England by the twelfth century. Tower mills, where only the top part of the windmill carrying the sails rotated, were introduced around the fourteenth century in France.

The industrial revolution brought refinements to windmill design but the number of mills began to decline with the advent of steam power. Even so there were still heavy concentrations of traditional mills in the Netherlands, where the Zaan district still boasted 900 windmills in the nineteenth century.

A new use for wind power developed with the invention of the wind pump, first used extensively among farmers in the USA. From the middle of the nineteenth century onwards, the classic lattice metal tower carrying a rotor with petal-shaped iron vanes crossed America and then travelled the world.

During the early part of the twentieth century there were several important experiments with the use of the wind to generate electricity, particularly in the USA and in Denmark. These failed to attract widespread attention. Finally, it was during the oil crisis of the early 1970s that modern interest in wind turbines took form. The main centres of development were the USA, particularly California, and Denmark.

From this period the basic wind energy conversion system for power generation has gradually taken shape. Today the basic system starts with a large rotor comprising two, three or four blades mounted on a horizontal shaft at the top of a tall tower. The blades intersect the wind and capture the energy it contains, energy which causes them to rotate in a vertical plane about the shaft axis. The slow rotation of the shaft is normally increased by use of a gearbox, from which the rotational motion is delivered to a generator. The electrical output from the generator is then taken through cables down the turbine tower to a substation where the power is eventually fed into the electricity grid. The mechanical components at the top of the turbine tower – the rotor, gearbox and generator – are all mounted on a platform that can pivot, or yaw, about a vertical axis so that the rotor shaft is always aligned with the wind direction. Gearbox and generator and housed within a weather tight compartment called a *nacelle*.

Turbine size

The early wind turbines which were developed for power generation in the late 1970s and the early 1980s had generating capacities of around 30–60 kW. Hundreds of machines of this size were installed in wind farms in California.

Through the 1980s and into the 1990s, wind turbine capacities increased steadily. During the 1980s there were several pilot projects involving single wind turbines with capacities of over 1 MW but during the decade the standard turbine size tended to be between 300 and 500 kW.

During the 1990s, turbine unit size continued to increase steadily. By 1998, most new wind farms employed turbines with a capacity of between 600 and 750 kW. These modern, higher-output machines tended to provide greater efficiency than the smaller machines of the previous decade and the trend towards even larger machines continued. At the end of the 1990s, the typical wind turbine size had reached 1 MW.

As the new century dawned, manufacturers began to introduce a range of multi-megawatt machines. A unit of around 2 MW is the most common at the beginning of 2004 but larger machines are being installed and 5 MW machines are already being developed. These largest machines have blades up to 60 m in length, leading to rotor diameters of 120 m.

The largest machines are particularly popular for offshore developments where the high cost of a turbine foundation favours a large generating capacity. The end of the first decade of this century will almost certainly see wind turbines for offshore applications in the 6–10 MW range. It is not clear yet whether there is an ultimate limit to wind turbine size.

Horizontal or vertical?

As outlined above, the standard wind turbine has a vertical rotor attached to a horizontal shaft. This arrangement imposes certain restrictions on the wind turbine design. With a horizontal shaft, the rotor turns in a vertical plane and must be raised on a tower so that the blades are clear of the ground and of the turbulent layer of air next to it.

Gearbox and generator are attached directly to the turbine shaft so these, too, must be placed on the tower, high above ground. This raises the cost of both installation and maintenance. And a horizontal axis machine must include a yawing system so that the rotor and nacelle can be rotated as the wind direction changes.

There is an alternative, a vertical axis wind turbine. A vertical axis machine has all its weight supported by a ground-level bearing. Both gearbox and generator can also be placed on the ground, easing maintenance costs. And most designs for vertical axis wind turbines will operate with the wind blowing from any direction. A yawing system is unnecessary.

Several vertical axis designs have been tested in the past 30 years. The most exhaustively explored was the Darrieus wind turbine which is based on a design patented by a French engineer, G.J.M. Darrieus, in 1931. Often described as the eggbeater, this wind turbine design comprises a pair of thin, curved blades with an aerofoil cross section attached to a vertical shaft and looking very much like an eggbeater.

A number Darrieus wind turbines have been built and tested and the design achieved short-lived popularity during the 1980s. Developments

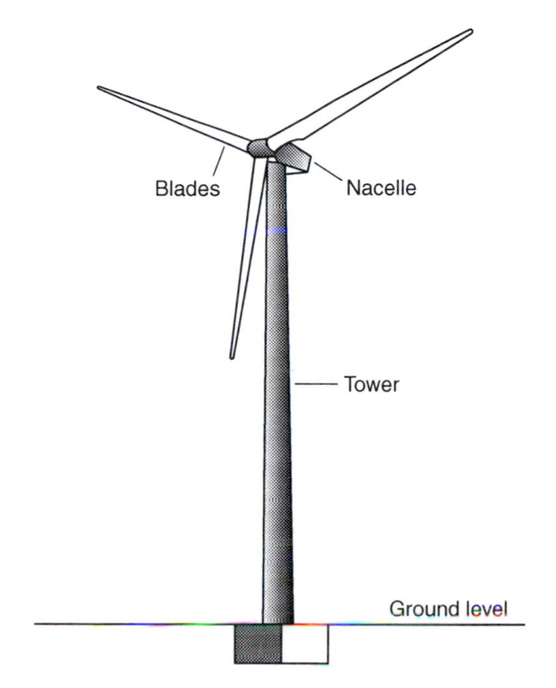

Figure 11.1 *A horizontal axis wind turbine*

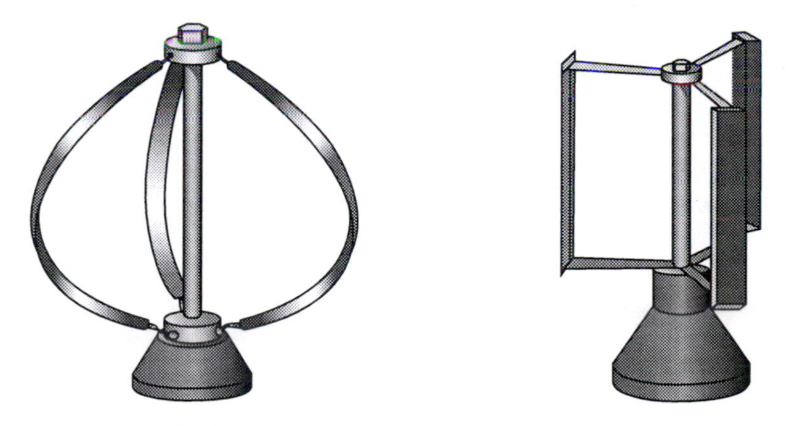

Figure 11.2 *A vertical axis wind turbine*

have included various pilot projects, a commercial machine and a prototype with an output of over 1 MW, developed in Canada. Other types of vertical axis turbine have also been built, including one with an H-shaped rotor which was operated at a test site in Wales. However the vertical configuration has not yet achieved significant commercial success.

Rotor design

Rotors for modern wind turbines come in a variety of guises. The chief variables are the number of blades and the means used to control the rotor speed in high winds.

Rotor blades can be made from a variety of materials including metal, wood, composite materials and carbon fibre. Metal blades were frequently used in early designs but are rarely seen in modern machines. Under modern design practice, low weight is considered a desirable property and the most common blade materials in use are glass-reinforced plastic and wood epoxy. Carbon fibre-reinforced epoxy resin is being introduced by some manufacturers. Its combination of strength and lightness are extremely attractive.

The rotor shape is aerodynamically determined. It also depends on the control method used to prevent the rotor turning too fast when wind speeds are high. One option is stall control. A stall-controlled rotor used fixed blades of a shape and orientation that causes the aerofoil to stall at high wind velocity, limiting its rotational speed. Stall-control machines tend to be more ruggedly built, and hence heavier than alternatives.

The second common option for speed control is pitch control. This involves a rotor with blades that can be twisted at different wind speeds to increase or decrease their aerodynamic effectiveness. Pitch control has an additional advantage of allowing some power-output control which can lead to better efficiency of operation. This system has become the most popular method of speed control.

Another design parameter is the number of blades on the rotor. Both weight and cost increase with the number of blades. Most current designs use three, though one, two and four blades have been tried.

A single-bladed machine offers the lowest-weight solution, but the single blade must be offset by a counterweight. As a result, two-bladed designs are often as light as the equivalent one-bladed machine when both rotor and nacelle are taken into account. The drawback with both one and two blades is an extra level of noise and although this can be reduced with design modification, a three-bladed rotor offers smoother rotational operation. Four blades offer good rotor balance but are heavier and less cost effective.

Rotor size is determined by the power output that the machine is designed to develop. The larger the diameter of the rotor, the more energy it can capture. A 600 kW wind turbine will have a rotor diameter of 40–50 m while a 5 MW machine needs a rotor diameter of around 120 m, rising to 140 m for 6 MW. Rotor sizes for offshore machines tend to be slightly larger than those of onshore machines for a similar output.

Rotational speed is also a factor in rotor design. It is normally necessary to keep the speed at which the tip of the rotor blades move through the air to below 70 m/s in order to minimise aerodynamic noise which can become an environmental problem. This means that as the rotor diameter increases, rotational speed must decrease. Current 2–3 MW designs normally rotate and between 5 and 20 rpm.

Tower design

Towers for wind turbines have been built in a variety of styles including steel lattice and cylindrical steel or concrete designs. The lattice style used to be common but has generally been replaced by cylindrical designs, predominantly of concrete but with steel used for some machines.

A tall tower will place the rotor in a region of high wind speed but will cost more. Thus tower height will normally be determined by the rotor diameter and by the need to avoid the layer of turbulent air close to the ground.

The bending frequency of the tower must also be taken into consideration. Excitation of this frequency could lead to structural damage in the same way as excitation of the natural frequency of a bridge can become dangerous.

Drive train and generator

As already noted, a modern wind turbine rotates at between 5 and 20 rpm while conventional generators operating at between 800 and 3600 rpm, so some form of step-up gearbox is usually necessary. This gearbox has to be extremely rugged because it must be able to withstand more than simply the rotational force from the wind turbine rotor.

Since wind speed varies with height, the force on a turbine blade at the lowest point of its orbit will be less than at the highest point. This will create a constant bending force on the rotor which is transmitted through the shaft to the gearbox. The speed of the generator must also be controlled so that it remains synchronised with the grid. If the shaft feeding energy to the generator starts to rotate too fast there will be a countermanding force from the generator resisting this increase in speed. This will set up an additional rotational stress within the shaft. This force will also be felt within the gearbox.

To cope with this, a wind turbine gearbox needs to be a heavy-duty design. Even so the gearbox is the most likely component in a wind turbine to fail.

One solution to this problem is to eliminate the gearbox altogether and use a system where the rotor is connected directly to the generator. Direct drive generators capable of operating at the low speeds encountered in wind turbines are being developed but current designs are much heavier than more conventional generators.

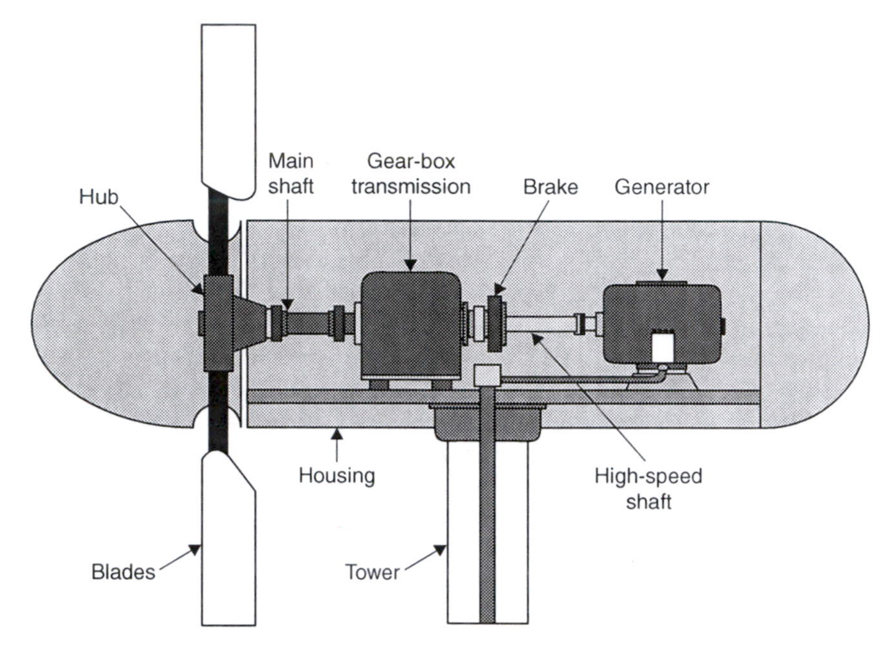

Figure 11.3 *A wind turbine drive train*

Another approach which eases the load on the drive train is to use a variable-speed generator. This will generate power at whatever speed the turbine rotor turns. However the variable frequency output must then be converted electronically to the grid frequency, adding to the cost of the wind turbine.

Wind farms and grid connection

To take full advantage of the wind, wind turbines are usually deployed in groups of from two or three to several hundred. These groupings are commonly known as *wind farms*.

When grouped together, wind turbines are usually spaced between five and ten rotor diameters apart in order to reduce interaction between adjacent machines. Even so, when machines are operating downwind of one another there will usually be some loss of output from the downwind turbines.

When this spacing is taken into account, a wind farm of twenty 500 kW turbines will occupy an area of 3–4 km^2. Of this, only around 1% is actually taken up by the turbines. The remainder can still be used as farmland.

The power from a wind farm must be delivered to the local grid. This will normally require a substation. For a small wind farm, under 100 MW, connection may be made to the local distribution system. Larger facilities, such as offshore farms, can have capacities of several hundred megawatts, possibly larger. These must be connected to the high-voltage transmission system.

Small wind farms have often relied on the grid to provide them with frequency and voltage stabilisation. However this puts a significant strain on the grid system which cannot support a large wind generation capacity in this way. Future wind farms will need to provide their own frequency and voltage stabilisation. This can be achieved today with standard techniques. However for the future electronic power-conditioning systems which convert the wind farm output to DC and then back to AC at the grid frequency and voltage will probably offer the most stable connection. This will only be cost effective for large wind farms.

Offshore wind technology

Pressures for land use and concerted campaigns to prevent the construction of wind farms is forcing wind farm developers in western Europe to consider building wind farms offshore. Offshore wind farming has some significant advantages. The wind regime is both more predictable and more reliable. Turbulence is lower, so wind turbines should last longer and wind farms can be sited far enough offshore to make them virtually invisible. Offshore sites also offer the possibility of building wind farms with capacities of 1000 MW or more.

Against this, the primary barrier is cost. Building a wind farm offshore costs between 40% and 100% more than building a similar farm onshore. Maintenance costs are higher too. However the higher wind speeds available offshore mean that output will generally be higher offshore.

The main additional cost is for construction of the wind turbine foundation. This can cost up to 25% of the total installation cost offshore. Onshore it is likely to be 16% or less. Grid connection is also more expensive. As a result, while the turbine itself may account for 64% of the cost of an onshore installation, it can be 45% less of the total offshore cost.

This high foundation cost favours large wind turbines for offshore projects. In 2004, the typical offshore wind turbine was between 2 and 4 MW in capacity and larger offshore turbines were in development.

A variety of methods have been used to build offshore foundations. The most popular is a monopile, a single pillar support which is constructed using either pile driving or drilling. Tripods have also been used and experimental work is examining the plausibility of using a floating wind turbine support.

Conditions offshore are generally more severe than onshore and offshore turbines must be 'marinised' to protect them from the environment. This adds around 10% to the cost of the offshore wind turbine compared to onshore. Marinisation techniques from the offshore oil and gas industry have been exploited to protect offshore turbines. The experience from this industry is also proving valuable when installing offshore wind turbines.

Constraints on wind capacity

The amount of wind power that can actually be exploited is likely to be limited, eventually, by the amount of wind generated power national transmission systems can accept. The actual proportion remains a matter for debate but the Danish power industry has shown that it is possible to accept 20% wind energy without detrimentally affecting grid operation.

The problem lies in the fact that wind is not a reliable resource. The output from a wind turbine cannot be guaranteed from day to day. Over a wide area, some variability will be averaged out, but nevertheless a significant level of uncertainty will remain. Better wind forecasting will help reduce the short-term uncertainty but long-term fluctuations in wind output are unavoidable.

Faced with this uncertainty, power dispatchers cannot depend on wind for base-load generation. When wind power is available they can use it to displace other types of generation. When it is not, they must bring those other plants into service.

The situation can be alleviated somewhat by building additional storage capacity on the grid to absorb the fluctuations. Even so, there is a limit to the amount of uncertain power of this type that a system can support with ease. It is possible to operate wind farms with less uncertainty if they are rated more conservatively but that means operating at below full output for most of the time, a strategy which will increase overall cost.

Offshore wind farms present an additional problem because the power from the generating facility must be delivered to the national grid at a point on the coast. In general coastal grids are not particularly good places to make such a connection. The grid can be strengthened in order to make it capable of accepting a large input of power but this will raise the cost of offshore wind power.

Environmental considerations

The principal environmental advantage of wind power is that it is a renewable resource. This means that its exploitation does not lead to a depletion of a global natural resource in the way that the burning of coal or gas results in reduced reserves. As a consequence, wind power can contribute to a sustainable global energy future.

Wind power is also a clean source of energy. Its use does not lead to significant environmental or atmospheric emissions.

Table 11.3 contains estimates for the lifetime carbon dioxide emissions from a wind plant and from a coal- and a gas-fired power station. The lifetime assessment looks at emissions that take place during the manufacture of the components of a power plant as well as the emissions that take place during the whole of its operational service. As a result of the former, a wind

Table 11.3 *Lifetime missions of carbon dioxide for various power generation technologies*

	Carbon dioxide emissions (tonnes/GWh)
Coal	964
Gas	484
Wind	7

Source: Concerted action for offshore wind energy in Europe, Delft University, 2001.

plant releases 7 tonnes of carbon dioxide for each gigawatt-hour of power it generates. As the table shows, a coal plant releases well over 100 times more and a gas plant close to 70 times the amount from the wind plant.

Aside from carbon dioxide wind power produces less sulphur dioxide, less nitrogen oxides and less of the other atmospheric pollutants that are emitted by coal-fired power plants, and to a lesser extent by gas-fired plants. However wind power plants are not entirely benign. The use of wind power does have negative consequences for the environment. Key among these are visual impact and noise.

Visual impact usually attracts the most serious criticism. Wind farms cover a large area and they are impossible to hide. While actual land utilisation is low and the area occupied by a wind farm can be used for other purposes too, the sight of an array of wind turbines, often in otherwise undeveloped rural areas, is considered by many to be visually offensive.

The weight placed on the visual impact of a wind power development will vary from site to site and from community to community; this is a matter of taste and it is virtually impossible to quantify. Nevertheless it will restrict the available sites for wind power development.

The other major effect of a wind turbine is to generate noise. The noise, a low-frequency whirring, has been compared to the sound of wind in the branches of a tree but the constant frequency is likely to make it more intrusive than the sound of the wind. To this rotor noise must be added the mechanical noise emanating from the gearbox and generator and occasionally some electrical noise.

The blade noise is the most serious of these. Turbine noise is generally more intrusive when wind speeds are low but it will be masked by background noise provided the machine is far enough away from human habitation. This again will limit the possible sites for wind development.

Under certain circumstances wind turbines can also cause electromagnetic interference, affecting television reception or microwave transmission. This can normally be mitigated by simple remedial measures and by careful site selection.

The ecological impact of wind power normally centres on its effect on bird populations. The most obvious danger, of birds being injured or killed when flying through the rotor blades, appears on current evidence to offer only a small threat. This could be considered more serious if a colony of an endangered species lived in the vicinity of a proposed wind farm, but birds do seem to learn to take account of wind farms.

Site development could also damage fragile environments such as peat bogs. This should be avoidable if good construction practice is followed.

Offshore wind

The major objections to onshore wind farms are eliminated with well chosen offshore sites. These can be 15 km or more offshore, making them all but invisible. Beyond 45 km, the curvature of the earth should render a wind farm completely invisible from ground level. Noise is not usually a serious problem offshore either.

Offshore wind farms do create their own problems. They can interfere with fisheries and with shipping. Their construction will disrupt the seabed, though the effect of this appears to be temporary. There will be affects on marine life but these too appear to be small.

Offshore wind farms can also interfere with radar. This has led to a number of sites being vetoed by defence ministries on the basis of security. Ground-based radar is affected but airborne radar should not be, so more modern radar systems are generally less subject to interference.

Financial risks

Two primary sources of risk can be attached to wind power, a risk associated with the reliability of the wind power resource available at a particular site and the risk attached to the use of wind power equipment.

The wind resource risk, the risk that the wind will not blow as it was expected to, should be minimal provided and adequate feasibility study has been performed. While the strength of the wind on a particular day at a particular site cannot be predicted in advance, wind is normally reliable over longer periods. A windy site will not turn into a windless site, at least not over the lifetime of a wind power project; global climate change could lead to long-term changes in the wind regime.

At an operational level, improved wind forecasting will enable a wind farm operator to predict future output and this is likely to make the project more valuable to a transmission system operator. It is worth stressing again that the only way to control the wind resource risk is by carrying out a careful site study.

The other major source of risk is the wind turbine technology. Onshore wind farms have now reached a sufficient level of maturity that these risks are well understood. Wind turbines can be expected to operate for 20–30 years with availabilities of 95–98%. New technology is being introduced and this will be liable to an additional level of risk, but overall the risks must be considered manageable. Given the experience now available, onshore wind farms should be able to attract financial support without the imposition of onerous levels of interest.

Offshore wind is at an earlier stage in its development and the risks here are consequently higher. Long-term availability and the cost of maintaining offshore facilities still have to be established. However the technology is not significantly different to that employed onshore, so performance should be broadly predictable. Offshore wind will also benefit from a major construction programme in the UK, promoted by government renewable energy targets, which is expected to lead to the construction of several thousand megawatts of installed capacity. This should establish, by the end of the first decade of this century, the viability of offshore wind farming.

The cost of wind power

Ever since the modern wind power industry began to develop, the main question it has had to answer is the question of cost. Can wind power compete with conventional forms of power generation?

Early development in California during the 1980s was stimulated by government financial incentives; when these were dropped, the development of projects declined too. Californian wind development was also affected by the fall in the cost of oil that started in the late 1980s. Real oil prices fell by 75% between 1980 and 1992, according to the World Bank.

In Europe the development of wind power took off more slowly but during the middle of the 1990s it became well established with Denmark and Germany its most enthusiastic early supporters, followed by Spain. Here, again, however, government incentives have helped promote wind generation. The wind power market began to grow again in the USA at the end of the 1990s, encouraged both by incentives and by regulations which required utilities in some states to provide a proportion of their electricity from renewable sources.

Continuous development since the early 1980s has led to the cost of wind turbine installations falling rapidly during the 1980s and early 1990s. The World Bank estimated that wind technology costs fell by between 60% and 70% between 1985 and 1994. While prices are still falling, the rates are not so dramatic as they were. Current installation costs for an onshore wind farm at between €700/kW and €1000/kW. Offshore wind farms still cost around €1500/kW but this could drop to €1000/kW by 2010.

The installation cost is the main up-front cost of a wind farm. However energy costs also depend on the amount of wind available at a particular site. To this must be added the cost of financing the project. Operating costs must also be included before a final figure for the cost of each kilowatt-hour of electricity can be established.

Taking these factors into account, favourable estimates suggest that at the beginning of the twenty-first century modern onshore wind farms could generate electricity for €0.03/kWh at a wind speed of 10 m/s and €0.08/kWh at a wind speed of 5 m/s. Early commercial offshore wind farms generate power for between €0.05/kWh and €0.08/kWh. Generating costs have been predicted to fall by 36% between 2002 and 2010 and a further 24% between 2010 and 2020,[7] predictions which if borne out will make wind power more competitive still.

These figures imply that onshore wind is currently broadly competitive with coal-fired power generation but not with gas-fired generation. Less favourable reviews of wind power claim that it generates power for two to three times the cost of coal plants. Such reviews take account of changes needed to grid operation and the cost of strengthening transmission grids to support the input of power from regions that have previously been at the end of the supply chain.

In both cases, external costs of fossil fuel power generation, the costs attached to the effects of the atmospheric pollution these plants cause, are ignored. Such costs are difficult to estimate but a 1998 EU report put the external cost of coal-fired power generation between €0.018/kWh and €0.15/kWh while the external cost of gas-fired power generation was between €0.005/kWh and €0.035/kWh. The equivalent estimate for wind energy was from €0.001/kWh to €0.003/kWh.[8]

Adding these amounts to generation costs would make wind generated electricity relatively more competitive. Even without taking external costs into account, wind power will almost certainly become cheaper over the next 10–20 years while the cost of coal- and gas-fired generation is likely to rise.

While arguments about its cost effectiveness continue, in practice the future of wind power is likely to be determined by political decisions. Environmental concerns are increasingly leading to legislation which requires the introduction of renewable electricity generation. Aside from hydropower, wind power is the best placed renewable source to meet that need. If renewable energy is required, in many cases that renewable energy will be wind energy.

End notes

1 Global Potential for Wind Energy, A.J.M. van Wijk, J.P. Coelingh and W.C. Turkenburg, Proceedings of Amsterdam EWEC'91. The figures in

Table 11.1 are taken from Windforce 12, a study published by the European Wind Energy Association and Greenpeace in 2002.

2 IEA World Energy Outlook 2000, quoted in Windforce 12, a study by the European Wind Energy Association and Greenpeace published in 2002.

3 Windforce 12, a study by the European Wind Energy Association and Greenpeace published in 2002. The figures appear to be taken from University of Utrecht Study by Wijk and Coelingh, published in 1993.

4 Gridded State Maps of Wind Electric Potential, M.N. Schwartz, D.L. Elliott and G.L. Gower, presented at the 1992 American Wind Energy Association's Conference in Seattle.

5 Technical Offshore Wind Energy Potential Around the Globe, R. Leutz, T. Ackermann, A. Suzuki, A. Akisawa, T. Kashiwagi, Proceedings of European Wind Energy Conference, Copenhagen (July 2002).

6 European Wind Energy Association.

7 European Wind Energy Association, Windforce 12 report for Greenpeace, 2002.

8 Externalities of Fuel Cycles Extern E Report. Report No 10: National Implementation. DGXII, Joule, published by European Commission in 1998.

12 Geothermal power

Geothermal energy is the heat contained within the body of the earth. The origins of this heat are found in the formation of the earth from the consolidation of stellar gas and dust some 4 billion years ago. Radioactive decay within the earth continually generates additional heat which augments that already present.

The distance from the surface of the earth to its core is 6500 km. Here the temperature may be as high as 7000°C. As a result of the temperature gradient between the centre and the much cooler outer regions, heat flows continuously towards the surface. An estimated 100×10^{15} W of energy reaches the surface each year. Most of this heat cannot be exploited but in some places a geothermal anomaly creates a region of high temperature close to the surface. In such cases it may be possible to use the energy, either for heating or in some cases to generate electricity.

The region of the earth at the earth's surface is called its *crust*. The earth's crust is generally 5-km to 55-km thick. Starting from the ambient surface temperature, the temperature within the crust increases on average by 17–30°C for each kilometre below the surface. Below the crust is the mantle, a viscous semimolten rock which has a temperature of between 650°C and 1250°C. Inside the mantle is the core. The earth's core consists of a liquid outer core and a solid inner core where the highest temperatures are found.

Geothermal temperature anomalies occur where the molten magma in the mantle comes closer than normal to the surface. In such regions the temperature gradient within the rock may be 100°C/km, or more. Sometimes water can travel down through fractured rock and carry the heat back to the surface. Plumes of magma may rise to within 1–5 km of the surface and at the sites of volcanoes it actually reaches the surface from time to time. The magma also intrudes into the crust at the boundaries between the tectonic plates which make up the surface of the earth. These boundaries can be identified by earthquake regions such as the Pacific basin 'ring of fire'.

The most obvious signs of an exploitable geothermal resource are hot springs and geysers. These have been used by man for at least 10,000 years. Both the Romans and the ancient Chinese used hot springs for bathing and for therapeutic treatment. Such use continues in several parts of the world, particularly Iceland and Japan. A district heating system based on geothermal heat was inaugurated in Chaude-Aigues, France, in the fourteenth century; this system is still in existence.

Table 12.1 *Main geothermal users, worldwide*

	Capacity (MW)
USA	2850
Philippines	1850
Italy	770
Mexico	740
Indonesia	590
New Zealand	345
Iceland	140

Source: US Geothermal Education Office.

Industrial exploitation of hot springs dates from the discovery of boric acid in spring waters at Larderello in Italy around 1770. This led to the development of a chemical industry based on the springs. It was here, too, that the first experimental electricity generation based on geothermal heat took place in 1904. This led, in 1915, to a 250-kW power plant which exported power to the local region. Exploitation elsewhere had to wait until 1958 when a plant was built at Wairakei in New Zealand and the Geysers development in the USA which began in 1960.

Geothermal generating capacity has grown slowly since then. By the beginning of the twenty-first century there was roughly 8000 MW of installed geothermal capacity worldwide.[1] The largest user is the USA with around 2850 MW of installed capacity. The Philippines has 1850 MW while Italy has 770 MW, Mexico has 740 MW and Indonesia has 590 MW (see Table 12.1). In total 23 countries have exploited geothermal power but two, Greece and Argentina, no longer have operating capacity.

Geothermal energy is attractive for power generation because it is simple and relatively cheap to exploit. In the simplest case steam can be extracted from a borehole and used directly to drive a steam turbine. Such easily exploited geothermal resources are rare but others can be used with little more complexity. The virtual absence of atmospheric emissions means that geothermal energy is clean compared to fossil-fuel-fired power. The US Department of Energy classifies geothermal energy a renewable one.

The geothermal resource

There are three principle types of geothermal resource. The simplest to exploit is a source of hot underground water which either reaches the surface naturally or can be tapped by drilling boreholes. Where there is no underground water source, anomalies in the crust can create regions where the rock close to the surface is much hotter than usual. This hot rock can be accessed by drilling and though it is not practical to exploit the heat today, experimental work may make its use possible in the future. The third, and richest source, is the magma itself. This contains by far the greatest amount of heat energy but because of the temperatures and pressures found within it, this is also the most difficult geothermal energy source to exploit.

Estimating the amount of energy in the crust of the earth that could be exploited for power generation is not easy. It has been suggested that there is between 10 and 100 times as much heat energy available for power generation as there is energy recoverable from uranium and thorium in nuclear reactors. Certainly, the resource is enormous.

Geothermal fields

Geothermal fields are formed when water from rain or snow is able to seep through faults and cracks within rock, sometimes for several kilometres, to reach hot rock beneath the surface. As the water is heated it rises naturally back towards the surface by a process of convection and may appear there in the form of hot springs, geysers, fumaroles or hot mud holes.

Sometimes the route of the ascending water is blocked by an impermeable layer of rock. Under these conditions the hot water collects underground in the cracks and pores of the rock beneath the impermeable barrier. This water can reach a much higher temperature than the water which emerges at the surface naturally. Temperatures as high as 350°C have been found. Such geothermal reservoirs can be accessed by boring through the impermeable rock. Steam and hot water will then flow upwards under pressure and can be used at the surface.

Most of the geothermal fields that are known today have been identified by the presence of hot springs. In California, Italy, New Zealand and may other countries the presence of these springs led to prospecting usage of boreholes drilled deep into the earth to locate the underground reservoirs of hot water and steam. More recently geological exploration techniques have been used to try and locate underground geothermal fields where no hot springs exist. Sites in Imperial Valley in southern California have been found in this way.

Some geothermal fields produce simply steam, but these are rare. Larderello in Italy and the Geysers in California are the main fields of this type in use today though others exist in Mexico, Indonesia and Japan. More often the field will produce either a mixture of steam and hot water or hot water alone, often under high pressure. All three can be used to generate electricity.

Deep geothermal reservoirs may be 2 km or more below the surface. These can produce water with a temperature of 120–350°C. High-temperature reservoirs are the best for power generation. Shallower reservoirs may be as little as 100 m below the surface. These are cheaper and easier to access but the water they produce is cooler, often less then 150°C. This can still be used to generate electricity but is more often used for heating.

The fluid emerging from a geothermal reservoir, at a high temperature and usually under high pressure, contains enormous quantities of dissolved minerals such as silica, boric acid and metallic salts. Quantities of hydrogen sulphide and some carbon dioxide are often present too. The concentrated brine from a geothermal borehole is usually corrosive and if allowed to pollute local groundwater sources can become an environmental hazard. This problem can be avoided if the brine is re-injected into the geothermal reservoir after heat has been extracted from it.

Geothermal reservoirs are not limitless. They contain a finite amount of water and energy. As a consequence both can become depleted if over-exploited. When this happens either the pressure or the temperature – or both – of the fluid from the reservoir declines.

In theory the heat within a subterranean reservoir will be continuously replenished by the heat flow from below. This rate of replenishment may be as high as 1000 MW, though it is usually smaller. In practice geothermal plants have traditionally extracted the heat faster than it is replenished. Under these circumstances the temperature of the geothermal fluid falls and the practical life of the reservoir is limited.

Re-injection of brine after use helps maintain the fluid in a reservoir. However reservoirs such as the Geysers in the USA, where fluid exiting the boreholes is steam, have proved more difficult to maintain since the steam is generally not returned after use. This has led to a marked decline in the quantity of heat from the geysers. In an attempt to correct this, wastewater from local towns is now being re-injected into the reservoir. Some improvement has been noted.

Estimates for the practical lifetime of a geothermal reservoir vary. This is partly because it is extremely difficult to gauge the size of the reservoir. While some may become virtually exhausted over the lifetime of a power plant, around 30 years, others appear able to continue to supply energy for 100 years or more. Better understanding of the nature of the reservoirs and improved management will help maintain them for longer.

Brine–methane reservoirs

In some rare cases underwater reservoirs of hot brine are found to be saturated with methane. These reservoirs will normally occur in a region rich in fossil fuel. Where such as reservoir is found it is possible in principle to exploit both the heat in the brine and the dissolved methane gas to generate electricity. The only major reservoirs of this type known today are in the Gulf of Mexico.

Hot dry rock

Underground geothermal reservoirs are relatively rare. More normally hot underground rock is not permeated by water and so there is no medium naturally available to bring the heat energy to the surface.

Where hot rock exists close to the surface, it is possible to create a man-made hydrothermal source. This is accomplished by drilling down into the rock and then pumping water through the borehole into the subterranean rock. If water is pumped under high pressure it will cause the rock to fracture, creating faults and cracks through which it can move. (In fact underground rock often contains natural faults and fractures through which the water will percolate.) If a second borehole is drilled adjacent to the first, then water which has become heated as it has percolated through the rock can be extracted and used to generate electricity.

The first attempt at this hot-dry-rock technique was carried out by scientists from the Los Alamos laboratory in New Mexico in 1973. Since then experiments have been carried out in Japan, the UK, Germany and France. The most recent project is the European Hot Dry Rock Research project at Soulez-sous-Forets in France. Here boreholes have been drilled to 5 km below the surface and temperatures of 201°C have been found. The next stage of this project involves construction of a system with a power plant that can extract 30 MW of thermal energy from the hot rock.[2]

Estimates suggest that a hot-dry-rock system will need to provide 10–100 MW of energy over at least 20 years to be economical. The technology is still in an early stage of development and it is likely to be 10–20 years before commercial exploitation is possible.

Exploiting the magma

Extracting energy from accessible magma plumes which have formed within the earth's outer crust is the most difficult way of obtaining geothermal energy but it is also the most attractive because of the enormous quantities of heat available. A single plume can contain between 100,000 and 300,000 MW centuries of energy.

Drilling into, or close to such hot regions is difficult because the equipment can easily fail. As an additional hazard, if a drill causes a sudden release of pressure, the result can be explosive. And ways have yet to be found to tap the heat. Research continues but it is a long-term project with no immediate prospect of exploitation.

Location of geothermal resources

The easiest geothermal resources to exploit are those that can provide water or steam with a temperature above 200°C. Resources of this type are located almost exclusively along the boundaries between the earth's crustal plates, in regions where there is significant plate movement. These areas are found around the Pacific Ocean in New Zealand, Japan, Indonesia, the Philippines and the western coasts of North and South America, in the central and eastern parts of the Mediterranean, east Africa, the Azores and Iceland.[3]

Lower-temperature underground reservoirs exist in many other parts of the world and though these contain less energy they can be used to generate electricity too. A project installed in Austria in 2001, for example, generates electricity from 106°C water which is also used for district heating. However these reservoirs can be more difficult to locate in the absence of hot surface springs. Nevertheless there were around 60 countries using geothermal energy at the beginning of the twenty-first century for either heating, generating electricity or both.

Today it is difficult to estimate the size of this energy resource but as survey techniques improve, more accurate data will become available. Based on data available at the beginning of the twenty-first century, reservoirs located in the USA, for example, could provide 10% of the US electricity. The world geothermal resource based on underground reservoirs is probably larger than the combined size of coal, oil, gas and uranium reserves.

Hot underground rock is even more widespread and the amount of energy contained in these rocks is enormous. However its exploitation will require the development of hot-dry-rock technology. This technology is still in its early stages, as outlined above. Magma resources are also likely to be widespread but the extent of the resource has not been widely explored.

Geothermal energy conversion technology

There are three principle ways of converting geothermal energy into electricity. Each is designed to exploit a specific type of geothermal resource. The simplest situation occurs where a geothermal reservoir produces high-temperature dry steam alone. Under these circumstances it is possible to use a direct-steam power plant which is analogous to the power train of a steam turbine power station but with the boiler replaced by the geothermal steam source.

Most high-temperature geothermal fields produce not dry steam but a mixture of steam and hot water. This is most effectively exploited using a flash-steam geothermal plant. The flash process converts part of the hot, high-pressure liquid to steam and this steam, together with any extracted fluid directly from the borehole, is used to drive a steam turbine.

Where the geothermal resource is of a relatively low temperature a third system called a *binary plant* is more appropriate. This uses the lower-temperature geothermal fluid to vaporise a second low boiling point fluid contained in a separate, closed system. The vapour then drives a turbine which turns a generator to produce electricity.

Direct-steam power plant

Dry-steam geothermal reservoirs are extremely rare. Where they exist the steam, with a temperature of 180°C to 350°C, can be extracted from the reservoir through a borehole and fed directly into a steam turbine. Steam from several wells will normally be fed to a single turbine. The pipes which carry the steam from the wellheads to the turbine contain various filters to remove particles of rock and any steam which condenses en route.

The steam turbine in a direct-steam geothermal plant is usually a standard reaction turbine. Efficiency is low at around 30%. Unit size in modern plants is typically between 20 and 120 MW. In some cases the steam exiting the turbine may be released directly into the atmosphere. However the steam usually contains between 2% and 10% of other gases such as carbon dioxide and hydrogen sulphide. Under these circumstances the exhaust from the steam turbine must be condensed to remove the water and then treated to remove any pollutants such as hydrogen sulphide before release into the atmosphere. At the Geysers plant in the USA, sulphur is produced as a by-product of the treatment. Condensing the steam also extracts more energy, so increasing plant efficiency.

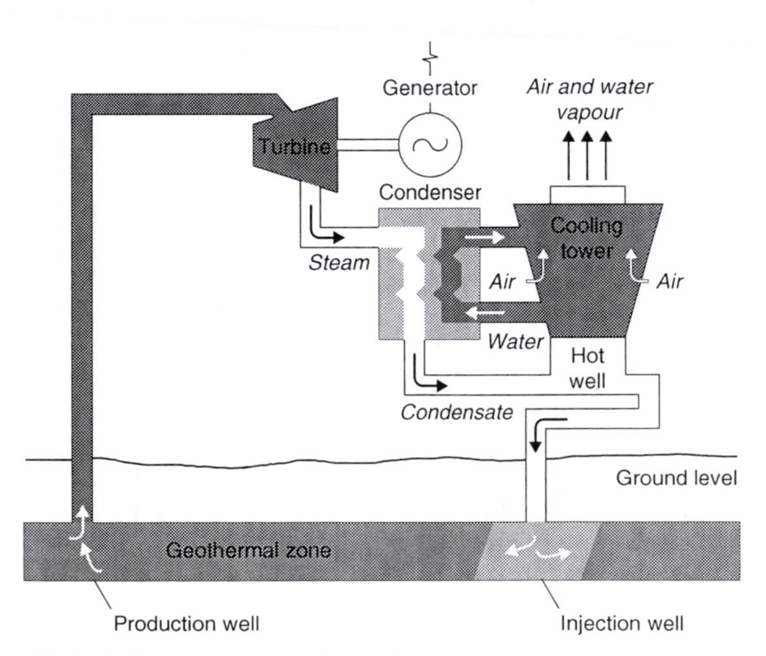

Figure 12.1 *A direct-steam geothermal power plant*

Ideally the geothermal fluid should be returned to the underground reservoir but it is often more economical to release the gas and dispose of the water produced as a result of condensing the steam from the turbine at the surface. Continual removal of fluid without replenishment eventually leads to a depletion in the quality of fluid available from the reservoir. At the Geysers geothermal field in southern California, urban wastewater is now being pumped into the underground reservoir in an attempt to maintain and eventually boost output from the resource.

Flash-steam plants

Most high-temperature geothermal reservoirs yield a fluid which is a mixture of steam and liquid brine, both under high pressure (typically up to 10 atmospheres). The steam content, by weight, is between 10% and 50%. The simplest method to exploit such a resource is to separate the steam from the liquid and use the steam alone to drive a steam turbine. However this throws away much of the available energy, particularly where the proportion of steam in the fluid is small.

A more productive alternative is to pass the combined fluid through a valve into a vessel maintained at a lower pressure than the geothermal fluid from the reservoir. The sudden reduction in pressure 'flashes' a proportion of the hot liquid to steam. This steam can then be separated from

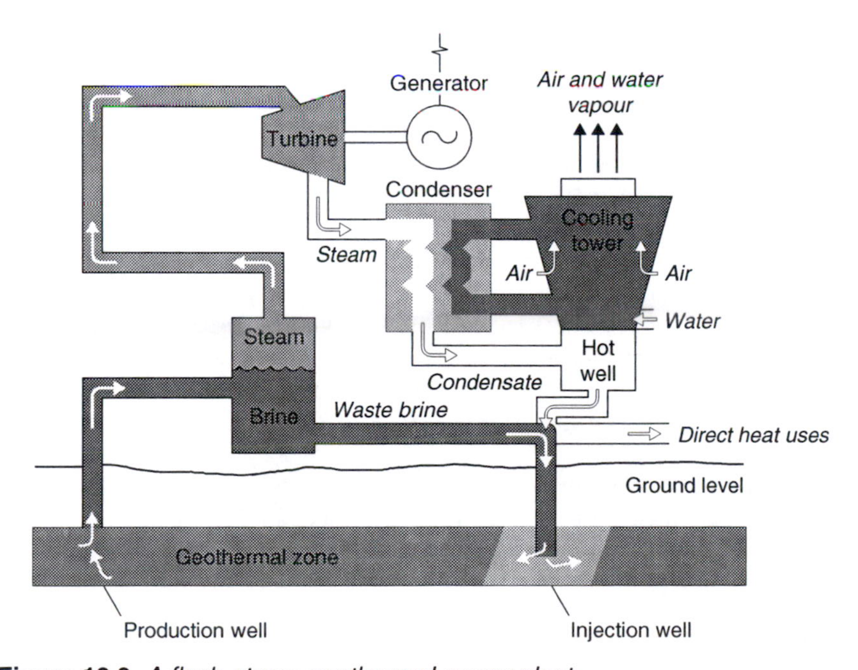

Figure 12.2 *A flash-steam geothermal power plant*

the liquid and used to drive a steam turbine. The steam exiting the steam turbine must be treated in exactly the same way as the exhaust from a direct-steam geothermal plant in order to prevent atmospheric pollution. The remaining liquid, meanwhile, must be injected into the geothermal reservoir since it usually contains high levels of dissolved salts which can cause pollution.

A further refinement to the flash-steam plant is called *double-flash technology*. This involves taking the fluid remaining from the first flash process and releasing it into a second vessel at even lower pressure. This results in the production of more steam which can be fed to a second, low-pressure turbine or injected into a later stage of the turbine powered by the steam from the first flash. A double-flash plant can produce up to 25% more power than a single flash plant. However it is more expensive and may not always be cost effective.

Flash technology plants will generally return a much higher percentage of the geothermal fluid – up to 85% for a single flash plant and somewhat less for the double-flash plant – to the geothermal reservoir. This will include most of the dissolved chemicals contained in the original fluid. However some reservoir depletion will still take place and without action this is likely to lead to a fall off in output from the reservoir with time. Capacities for flash geothermal power plants are normally between 20 and 55 MW.

Binary power plants

Direct- and flash-steam geothermal power plants utilise geothermal fluid with a temperature of between 180°C and 350°C. If the fluid is cooler than this, conventional steam technology will normally prove too inefficient to be economically viable. However energy can still be extracted from the fluid to generate power using a binary power plant.

In a binary power plant the geothermal fluid is extracted from the reservoir and immediately passed through a heat exchanger where the heat it contains is used to volatilise a secondary fluid. This secondary fluid is contained within a completely closed cycle system. The fluid may be an organic liquid which vaporises at a relatively low temperature or, in the case of the Kalina Cycle,[4] a mixture of water and ammonia.

The vaporised secondary fluid is used to drive a turbine from which power can be extracted with a generator. From the turbine the vapour is condensed and then pumped through the heat exchanger once more. Thus the cycle is repeated.

The geothermal fluid exiting the heat exchanger, meanwhile, is re-injected into the geothermal reservoir. Since 100% of the fluid is returned underground, this type of geothermal power plant has the smallest environmental impact. Typical binary plant unit size is 1–3 MW, much smaller than for the other types of geothermal technology. However the small

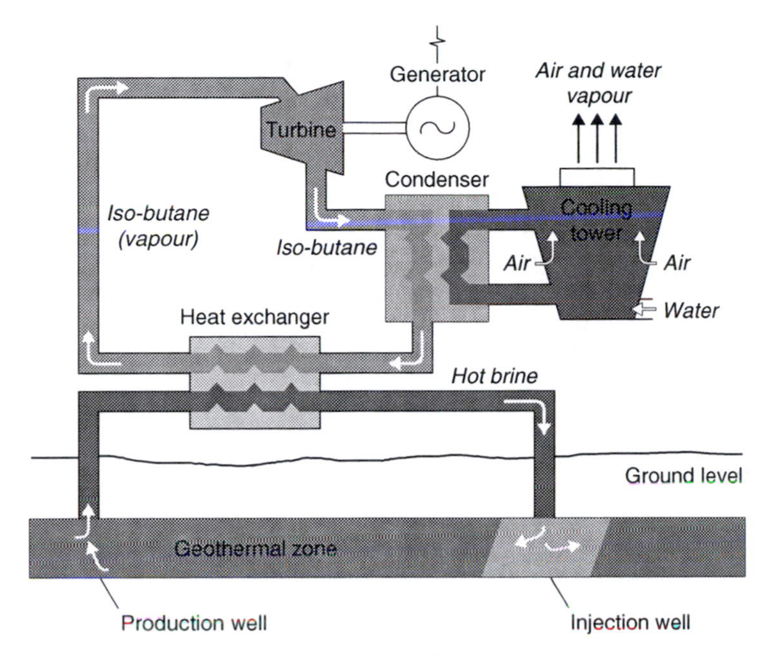

Figure 12.3 *A binary geothermal power plant*

modular units often lend themselves to standardisation, reducing production costs. Several units can be placed in parallel to provide a plant with a larger power output.

Although the normal application for binary technology is to exploit a low-temperature geothermal resource, the technology can also be used to generate more power from a flash plant. In this case the fluid left after flashing is passed through a heat exchanger before re-injection, allowing extra energy to be taken to power a small binary unit. Adding a binary unit to a conventional flash plant increases the cost but the resultant hybrid plant will have a larger power output.

Environmental considerations

Geothermal power generation is frequently classed among the renewable energy technologies. Strictly this is incorrect. Geothermal heat is mined from the earth and the heat removed to generate electricity is not replaced. Nevertheless the amount of heat contained within the earth is virtually limitless in human terms. For this reason, if for no other, geothermal energy can be treated as if it were renewable. As already noted above, this is the position of the US Department of Energy which classifies geothermal energy as renewable.

The construction of a geothermal power plant involves the same type of disruption that any civil engineering project entails. However to this is added the disruption associated with the drilling of wells to remove geothermal fluid from an underground reservoir and the re-injection wells to dispose of the fluid once it is exhausted of heat. Drilling requires significant quantities of water and this must be taken from local water courses. To minimise environmental effects, this should be taken from high-flow streams and rivers, preferably during the rainy season.

Geothermal resources are often associated with natural features such as fumaroles, geysers, hot springs and mud holes. These are natural features which will normally be protected by environmental legislation, so drilling directly into a reservoir that feeds such features will often not be possible. These features may also have social and religious significance which must be respected.

Management of the underground geothermal reservoir forms an important part of any geothermal project both on environmental and economic grounds. Continuous depletion of the reservoir will lead to a lowering of the local water table and may lead to subsidence as well as to a fall off in the quality of fluid from the boreholes. The quantity of fluid should be maintained by re-injection of all, or as much as possible of the extracted fluid. However re-injection can lead to a cooling of the reservoir. This can be avoided by carefully mapping the local flows and re-injecting some distance from the extraction site. Induced seismic activity has also been linked with re-injection, but a causal link is difficult to prove since most geothermal projects are in regions of high or regular seismic activity.

The emissions from a well-managed geothermal plant should be very small when compared with a conventional steam plant. Any carbon dioxide contained in the fluid from the subterranean reservoir will be released and there may be traces of hydrogen sulphide too. However the latter can be removed chemically to prevent release. The saline brine can cause serious groundwater pollution, as was experienced in New Zealand where the Wairakei power plant released 3500 tonnes/h of brine into the Waikato river. To prevent such pollution, modern geothermal plants re-inject all the extracted brine after use.

Financial risks

The major risk associated with developing a geothermal project concerns finding a geothermal resource suitable for exploitation. Initial surveys of geothermal resources are often carried out by national institutions but these will need to be backed up with test wells to determine the exact nature of the resource available. Data from Pacific rim volcanic regions suggests the presence of single hot spring will provide a 50% change of an exploitable geothermal field. A boiling spring or fumarole increases the probability to 70%.[5]

Having identified a suitable surface site, pre-feasibility studies are likely to cost around $1 million, with a 30% change of failure. Test drilling, usually three wells at up to $2 million per well, has a similar prospect of failure. This risk can be reduced by careful surface study followed by prioritisation of the available sites. Such an approach has led to success rates for well drilling in excess of 83% in countries such as Indonesia, Kenya and New Zealand which have high-temperature resources. However success rates can be much lower where low-temperature resources are concerned.

Once a usable underground reservoir has been located, its size must be determined. This involves fluid withdrawal over a long period; indeed it may not be until several years after production has started that a good picture of the resource can be obtained.

Careful sizing of the geothermal plant to match the reservoir size will prolong the lifetime of a reservoir. This may not be possible if the plant has to be constructed before full data is available. Oversized plants such as that installed at the Geysers in the USA lead to a premature fall in output.

Having identified a reservoir and assessed its potential, the risk associated with the power plant technology used to exploit it is minimal. All geothermal technologies in current use are well tested and predictable.

New methods of exploiting the heat energy in the earth such as hot-rock techniques are still in an early stage of development and the risks here are large. However this technology is a long way from commercial exploitation.

The cost of geothermal power

In common with many renewable resources, geothermal power generation involves a high initial outlay but extremely low fuel costs. In the case of a geothermal plant there are three initial areas of outlay, prospecting and exploration of the geothermal resource, development of the steam field and the cost of the power plant itself.

Prospecting and exploration may cost as much as $1 million. This will weigh more heavily on small geothermal projects than on larger schemes. Steam field development will depend on plant size, as will the cost of the power plant itself, though small plants tend to be more expensive than larger plants.

Table 12.2 shows figures from the World Bank for the costs of development of geothermal projects for different qualities of geothermal resource. A good resource has a temperature above 250°C, and good permeability so providing good fluid flow. It will provide either dry steam or steam and brine, with low gas content and the brine will be relatively non-corrosive. A poor resource may have a temperature of below 150°C, but it could provide fluid at a higher temperature with some other defect such as a corrosive brine or poor fluid flow.

Table 12.2 *Direct capital costs ($/kW) for geothermal power plants*

Plant size (MW)	*Resource*		
	Good	*Medium*	*Poor*
<5	1600–2300	1800–3000	2000–3700
5–30	1300–2100	1600–2500	–*
>30	1150–1750	1350–2200	–*

*Not usually suitable.
Source: World Bank.

As the figures in the table show, costs for a good resource vary between $1150 and $2300/kW depending on plant size. Where the resource is poor, large plants are not normally economically viable. Costs for small power plants under these circumstances vary between $2000 and $3700/kW.

Further indirect costs will be incurred, depending on the location and ease of access of the site. These will vary from 5% for an easily accessible site and a local skilled workforce to 60% of the direct cost in remote regions where skilled labour is scarce. An alternative cost estimate from the US Energy Information Administration put the cost of a 50 MW geothermal power plant entering service in the USA in 2006 at $1700–1.800/kW.

These costs will all be part of the initial investment required to construct a plant. Electricity production costs will depend partly on this, partly on financial arrangements such as loan repayments and partly on continual operation and maintenance costs. World Bank estimates suggest that power can be produced from a large geothermal power plant (>30 MW) exploiting a good quality resource at between $0.025 and $0.050/kWh. A plant of less than 5 MW could generate power from a similar resource for $0.050–0.070/kWh. With a poor quality resource a small geothermal plant can generate for $0.060–0.150/kWh.

Based on these estimated, a large geothermal power plant can hope to compete with gas-fired power plants. Small plants are less economical but they can still offer extremely competitive power in remote rural areas where the alternative is a diesel power plant. Power from the latter will cost at least $0.10/kWh, probably much higher.[6]

End notes

1 World Status of Geothermal Energy Use; Overview 1995–1999, John W Lund, Geo-Heat Center, Oregon Institute of Technology, Klamath Falls, OR, USA.

2 Development of Hot Dry Rock Technology, Helmut Tenzer, GHC Bulletin, December 2001, p14.
3 World Energy Council, Survey of Energy Resources, 2001.
4 The Kalina cycle is a special thermodynamic cycle designed to obtain maximum efficiency from low-energy resources such as low-temperature geothermal fluids.
5 Geothermal Energy, an Assessment, World Bank, available at www.worldbank.org
6 Refer *supra* note 5.

13 Solar power

Solar energy is the most important source of energy available to the earth and its inhabitants. Without it there would be no life. It is the energy source that drives the photosynthesis reaction. As such, it is responsible for all the biomass on the surface of the earth and is the origin of fossil fuels, the products of photosynthesis millions of years ago and now buried beneath the earth's surface. Solar energy creates the world's winds, it evaporates the water which is responsible for rain; waves and ocean thermal power are both a result of insolation. In fact, apart from nuclear energy, geothermal energy and tidal power, the sun is responsible for all the forms of energy which are exploited by man.

All these different sources of energy, each derived from the sun, can be used to generate electricity. However solar energy can also be used directly to generate electricity. This can be achieved most simply by exploiting the heat contained in the sun's radiation, but electricity can also be generated directly from light using an electronic device called a *solar cell*. Both methods are valuable renewable sources of electricity.

The solar energy resource

The energy radiated by the sun is around 7% ultraviolet light, 47% visible light and 46% infrared light. Its energy content at the distance of the earth from the sun is around 1.4 kW/m^2. Each year around 1500 million TWh of solar energy reaches the earth.

Not all this energy reaches the surface of the earth. Much of the shorter wavelength ultraviolet radiation is absorbed in the atmosphere. Water vapour and carbon dioxide absorb longer wavelength energy while dust particles scatter more radiation, dispersing some of it back into space. Clouds also reflect light back into space.

When all these factors are taken into account, around 47% of the energy, 700 million TWh actually reaches the surface. This is 14,000 times the amount of energy, 50,000 TWh, used by mankind each year. Much of this solar energy strikes the world's oceans and is inaccessible. Even so, with reasonably efficient energy conversion systems, less than 1% of the world's land area would provide sufficient energy to meet global electricity demand, around 15,000 TWh.

Let us put this into a more practical perspective. A group of solar thermal power plants were built in California in the late 1980s and early 1990s. The design of these plants was based on an estimated solar energy input of 2725 kWh/m^2/year. This is equivalent to 22.75 GWh for each hectare each year. On this basis, assuming a conversion efficiency of 10%,[1] 10 million ha, or 100,000 km^2 (316 km $\times$ 316 km), could generate enough energy to supply the entire USA.

This may appear to be a large area but the demand is not onerous. Such an area could be found quite easily, particularly if desert areas were exploited. In fact solar electricity generation takes up less land than most hydropower projects where these include reservoirs. Indeed the land requirements of some hydro schemes can be as much as 50 times a typical solar project yielding the same output.[2]

But in spite of its enormous potential, global solar electricity generating capacity is tiny. According to European Union estimates, there was probably less than 800 MW of installed capacity in 1995 (including all types of solar generation technologies). Between 1995 and the end of 2003, gross world production of solar cells was around 2600 MW. With little other additional solar capacity, total global solar generating capacity may have been 3400 MW at the beginning of 2004.

Sites for solar power generation

In principle solar power can be generated anywhere on the earth but some regions are better than others. Places where the sun shines frequently and regularly are preferable to regions where cloud cover is common. The brighter the sunlight, the greater the output and the more advantageous the economics of the generating plant. Many of the world's developing countries, where demand for electricity is growing rapidly, offer good conditions for solar electricity generation.

Solar generating stations do not take up enormous amounts of land but they do require many times the space of a similarly sized fossil fuel power plant. But solar power does not necessarily require large contiguous areas of land in order to generate electricity. Solar panels can be made in small modular units which can be incorporated into buildings so that power generation can share space used for other purposes.

Distributed generation of this type has many advantages. In California, and elsewhere, there is a major daytime grid demand peak resulting from the use of air-conditioning systems. As the air-conditioning systems are used to combat heat generated by the sun, distributed solar electricity generation matches this demand perfectly. Recent experience has shown that domestic solar panels virtually eliminate this additional demand from the houses to which they are fitted.[3]

Solar technology

There are two ways of turning the energy contained in sunlight into electricity. The first, called *solar thermal generation*, involves using the sun simply as a source of heat. This heat is captured, concentrated and used to drive a heat engine. The heat engine may be a conventional steam turbine, in which case the heat will be used to generate steam, but it could also be a gas turbine or a sterling engine.

The second way of capturing solar energy and converting it into electricity involves use of the *photovoltaic* or *solar cell*. The solar cell is a solid-state device like a transistor or microchip. It uses the physical characteristics of a semiconductor such as silicon to turn the sunlight directly into electricity.

The simplicity of the solar cell makes it an extremely attractive method of generating electricity. However the manufacture of the silicon required for solar cells is energy intensive. The solar thermal plant, although more complex is currently cheaper and uses more conventional power station technology.

Whatever its type, a solar power plant has a major weakness. It can only generate electricity when the sun is shining. During the night there is no sunlight and so no electricity. In order to circumvent this problem, a solar power station must either have some form of conventional fuel back-up, or it must incorporate energy storage. Solar cells are frequently coupled with rechargeable batteries in order to provide continuous power in remote locations. Solar thermal power plants can also be designed with heat storage systems which allow them to supply power in the absence of the sun.

Solar thermal power generation

The sun is a source of high-quality heat which can easily be exploited for power generation. This was recognised as early as 1907 when the first patent for a solar collector was granted in Germany to Dr W. Maier. The development of modern solar thermal power technology began in the 1970s and was finally proved in the late 1980s with a series of commercial solar thermal power plants in California.

In spite of the success of these plants no further commercial plants have been built anywhere in the world. Research has continued, however, and interest accelerated at the beginning of the twenty-first century following renewed support from government and international agencies. It seems likely that several new projects will be built before the end of the first decade of the new century.

Modern solar thermal research has concentrated on three different approaches to converting solar energy into electricity. All require sunlight to be collected and concentrated to provide a high-energy source. The first uses a parabolic trough-shaped mirror to focus the energy contained in sunlight onto an energy collector at the focus of the parabola. These parabolic trough solar units can be deployed in massive arrays to provide a large generating capacity.

The second approach, called a *solar tower*, employs a solar energy collector mounted atop a large tower. A field of mirrors is used to direct sunlight onto the collector where the concentrated heat is used in a power generation system. Both this and the parabolic trough system can be used to build utility-sized power plants. The third system, usually called the *solar dish*, comprises a parabolic dish with a solar heat engine mounted at its focus. Dishes are usually only 10–50 kW in capacity but can achieve high-energy conversion efficiency. Fields of dishes are needed to produce a high-capacity power plant.

There is also a novel technique being explored in Australia called a *solar chimney*. This involves building a massive greenhouse, in the centre of which is an extremely tall chimney. The chimney sucks hot air from the greenhouse, creating a massive updraft. Fans or turbines placed inside the chimney can capture energy from this updraft to generate electricity. It is estimated that 40 km^2 of greenhouse and a chimney 1000 m high will be needed to generate 200 kW.

Parabolic troughs

The sunlight which reaches the earth, while it can feel extremely hot, does not contain sufficient energy in the diffuse form in which it arrives to constitute the basis for a thermal power generation system. In order to make it useful, the sunlight from a large area must be concentrated. This can be achieved with a magnifying lens but lenses form a relatively expensive way of concentrating sunlight. Much better is a concentrating reflector.

The parabola is the ideal shape for a solar reflector because it concentrates all the light incident on it from the sun at a single point called the *focus*. A complete parabola is circular; this forms the basis for solar dish system (see below). However there is a limit to the size of dish which can be built. For large-scale solar concentration, a trough-shaped reflector has proved more effective. If the trough is built with a parabolic cross-section, the reflector will bring the incident sunlight to focus at a line rather than at a single point, a line running along the length of the trough. This is the basis for the solar trough.

The reflecting panels in a solar trough are made from mirrored glass, although cheaper options are also being developed. These reflecting panels are mounted on a substantial substructure capable of supporting their weight. Along the length of each trough, mounted exactly at the focus of the parabolic cross-section, is a heat absorbing tube. This is where the solar heat is collected.

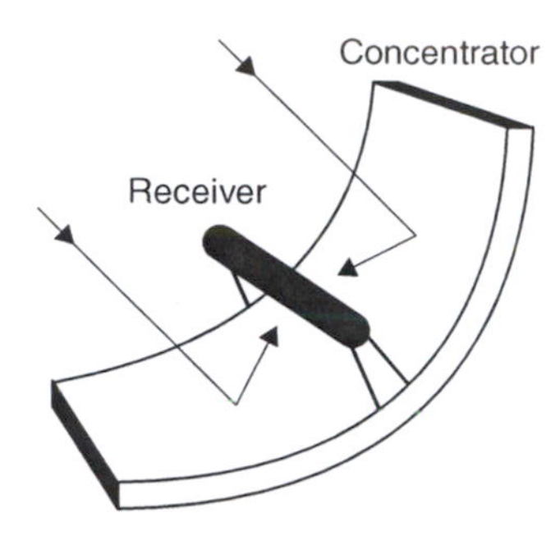

Figure 13.1 *A solar trough system*

An individual solar trough reflector may be up to 150 m in length. Arrays of parallel troughs provide the required collection and generating capacity. A large number or troughs will be required to build a utility-sized power station. For example, a single 30 MW power plant in California employs 980 parabolic trough collectors, each 47 m long.

In order to achieve the highest efficiency, the solar troughs should track the sun across the sky. If the solar troughs are aligned north–south, a system that tilts the troughs about their long axes can be used to follow the sun from east to west. This is the arrangement which has been used in existing solar trough power plants.

Once sunlight has been concentrated it must be captured and converted into a form of energy that can be used to generate electricity. The simplest way of achieving this is to place a tube containing a heat absorbing liquid at the focus of the parabolic trough. The liquid is pumped through the tube, absorbing heat as it passes and this heat is used to provide energy to drive a heat engine.

The small number of commercial solar trough power plants in operation all use a heat absorbing oil as the heat collection and transfer medium. This is pumped through the absorber tubes in the solar troughs, where it eventually reaches close to 400°C. The hot oil is then pumped through a heat exchanger where it is used to heat water and raise steam in a secondary system. The steam drives a steam turbine which generates power.

Commercial experience with parabolic trough solar power is based on nine plants in California built from the end of the 1980s with capacities ranging from 3.8 to 80 MW. These plants employed a secondary firing system utilising natural gas so that output could be maintained when solar input was low. In total the fossil fuel input could provide up to 25% of the plant output. In 2000, these plants achieved a peak solar-to-electric energy conversion efficiency of 23% and an annual efficiency of 15%.

The nine Californian plants have proved a success, commercially, but they were expensive to construct and would be unlikely to attract support

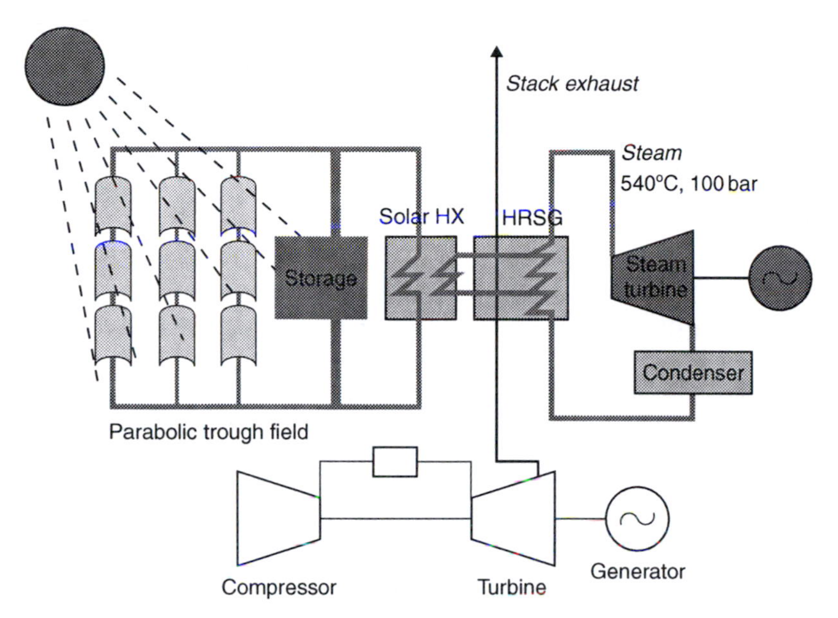

Figure 13.2 *An integrated-solar-thermal/combined cycle power plant utilising solar troughs*

in today's deregulated electricity market. However new designs are expected to become competitive. Research suggests that costs can be reduced significantly if new reflector fabrication and mounting methods are developed. Another significant cost reduction can be achieved by dispensing with the intermediate heat transfer oil used in existing plants and heating water directly to generate steam. Some form of heat storage system will improve flexibility and economics by allowing solar plants to generate power at night.

Another approach is to build on the idea of a combination of solar and fossil fuel energy by developing sophisticated hybrid power plants. The most attractive scheme involves building a parabolic trough collector array adjacent to a combined cycle power plant and using the heat collected by the solar array to supplement the heat from the gas turbine exhaust of the combined cycle plant. Both the gas turbine exhaust heat and the solar heat are then used to raise steam and drive a steam turbine.

This arrangement, called an *integrated-solar combined cycle* (ISCC) plant, makes good use of solar energy when it is available but is not reliant on the sun. Incorporating solar collection into a conventional fossil fuel power plant reduces the cost of the solar energy system significantly and this approach has attracted the support of the World Bank's Global Environment Facility. Such plants might involve a solar generating capacity of 30–40 MW out of a total of 140–300 MW. While the ISCC plant still relies primarily on fossil fuel, it does represent an economical way of introducing significant solar capacity.

Solar towers

The solar tower takes a slightly different approach to solar thermal power generation. Whereas the parabolic trough array uses a heat collection system spread throughout the array, the solar tower concentrates heat collection and utilisation at a single central facility.

The central facility includes a large solar energy receiver and heat collector which is fitted to the top of a tower. The tower is positioned in the centre of a field of special mirrors called *heliostats*, each of which is controlled to focus the sunlight that reaches it onto the tower-mounted solar receiver.

The mirrors used as heliostats must be parabolic in section, just like the trough mirrors, but because they have a very long focal length they appear almost flat. Each mirror has to be able to track the sun independently so that the incident light remains directed at the solar receiver. The heliostat field can be very large, large enough to supply energy to generate several hundred megawatts of electricity.

The most technically demanding component of a solar tower system is the heat capture and transfer system. At the top of the solar tower is a solar receiver containing tubes through which a heat transfer fluid flows. This has to be capable of absorbing the heat from the whole heliostat field. Once heated, the fluid is pumped to a heat exchanger where the heat is used to generate steam for a steam turbine.

This arrangement is much like the parabolic trough power plant, but modern designs include a crucial difference. The heat transfer fluid in a solar tower is not pumped directly from the solar receiver to the heat exchanger. Instead it is taken to a high-temperature storage tank, a heavily insulated tank where the hot fluid can be stored for up to 24 h. From here the fluid is taken as needed and pumped through the heat exchanger to generate steam, and then it is stored in a low-temperature reservoir. Fluid from this low-temperature reservoir is supplied to the solar receiver to be reheated.

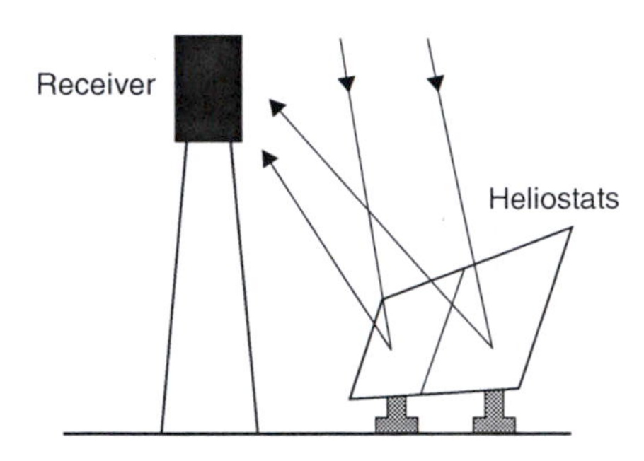

Figure 13.3 *A solar tower system*

By careful sizing of the storage system and power generation system, a solar tower can be constructed so that it can supply power continuously, not just during daylight hours. This means that the plant can be employed like a normal base-load fossil-fuel-fired power station, making it much more flexible and therefore much more valuable on a grid system.

A key component of the solar tower is the heat transfer fluid. The most successful has proved to be a molten salt comprising a mixture of sodium and potassium nitrates. This will melt at 220°C. It is normally kept at around 300°C in the low-temperature reservoir and is heated to 550°C in the solar receiver for storage in the high-temperature reservoir.

No commercial solar tower power plants have been built but a number of pilot-scale projects have been operated. Most important of these have been two projects at Barstow in California, called Solar One and Solar Two. Solar One operated from 1982 until 1988. It was later cannibalised to build Solar Two which started up in 1996 and operated until 1999. Both had power generating capacities of 10 MW.

Solar One used water as its heat transfer medium but it was converted to a molten salt system with two storage tanks for Solar Two. The latter comprised a 91-m high tower surrounded by 2000 heliostats with tracking systems which were computer controlled. At the top of the tower was the solar receiver, a system of vertical pipes which carried a molten salt. This molten salt reached a temperature of 565°C when heated by the sun. Storage capacity was 30,000 kWh.

The solar tower is a source of very high-grade heat. While pilot plants have operated with temperatures of around 550°C it is quite plausible to raise the temperature to 1000°C. Such a high temperature could be used to heat air to drive a gas turbine, instead of for raising steam. A gas-turbine-based system could prove more efficient than the current steam turbine solar tower design, but the scheme has yet to be proved. There may also be ways of combining a solar tower with a fossil fuel power plant in a hybrid arrangement similar to that being considered for the ISCC power plant described above.

The solar tower concept has never been proved commercially but it is considered to be perhaps the most cost effective of the solar thermal technologies. However it will be the second or third decades of the twenty-first century before it reaches commercial maturity.

Solar dish collectors

The third type of solar thermal power unit is the solar dish. A solar dish is more accurately a parabolic mirror, at the centre of which is placed a small heat collector and electricity generator. The reflector tracks the sun and focuses its energy onto the collector.

Unlike the two preceding technologies which are being developed for utility scale generation, the solar disk will always be a relatively small-scale

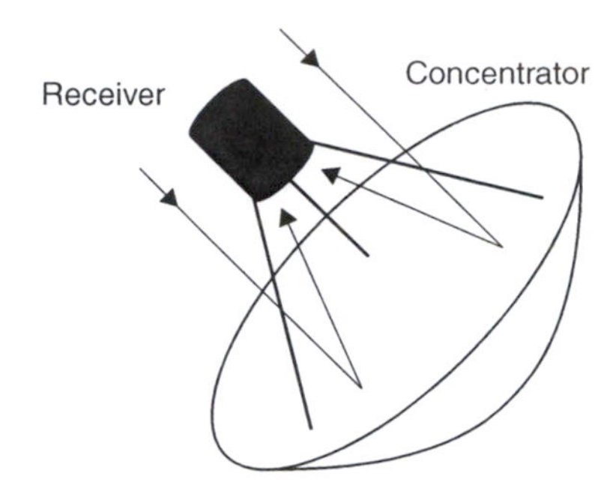

Figure 13.4 *A solar dish*

electricity plant. Those currently being tested have diameters of between 5 and 15 m and outputs of 5–50 kW. Larger dishes seem possible and there is a plan to build one with a capacity of greater than 1 MW, but even so utility multi-megawatt capacities can only be achieved by installing large numbers of individual units.

The two key components of a dish system are the parabolic reflector and the heat engine. Since the reflector must track the sun, a tracking system must also be included. Reflectors can be made using traditional glass-based techniques but these are very heavy and new, lighter fabrication methods are needed to bring down costs.

The most popular type of engine for use with a solar dish is a sterling engine. This is a piston engine (see Chapter 6) but a piston engine in which the pistons are part of a completely closed system. The energy source, heat, is applied externally. Consequently this is perfectly suited to solar dish applications.

The solar dish is the most efficient of all the solar thermal technologies. The best recorded solar-to-electrical conversion efficiency is 30%, but the Stirling engine is theoretically capable of 40% efficiency. This is of importance because of the area needed for a solar power plant. While parabolic trough systems require 2.2–3.4 ha for each megawatt of generating capacity, solar dishes need 1.2–1.6 ha.[4]

Solar dishes are currently expensive but costs can be reduced significantly. However they are unlikely to be a cost effective as the solar tower. Their main use is likely to be for stand-alone remote generation where their high efficiency and reliability could eventually challenge that of the solar cell, the solar device currently used most widely for such applications.

Photovoltaic devices

The solar photovoltaic device, more commonly known as the *solar cell*, exploits a completely different means of converting sunlight into electricity. This depends on the physical characteristics of materials called *semiconductors*.

The solar cell is a solid-state device which shares a heritage with the diode, the transistor and the microchip. It was developed in the Bell Laboratories in the early 1950s and soon found action in the US space programme. Today it remains the most widely used means of providing electric power to satellites and space vehicles.

Solar cells began to be used for terrestrial applications during the 1980s, mainly in remote locations where reliable power was needed without regular human intervention. As costs began to fall (although they remained extremely expensive) their use was extended to a wider range of applications. From 1990 onwards, grid-connected solar cells began to appear in domestic and some commercial applications. This usage continued to expand during the first years of the twenty-first century.

Solar photovoltaic technology

The solar cell is made from a thin layer of semiconducting material. The key feature of this semiconductor layer is that it will absorb photons of radiation in the visible region of the electromagnetic spectrum. Each photon of light energy is absorbed by an electron within the solid material. In absorbing the energy, the electron acquires an electrical potential. This potential can be made available as electrical energy, as an electric current. The current is produced at a specific fixed voltage called the *cell voltage*. The cell voltage is a property of the semiconducting material. For silicon it is around 0.6 V.

The energy contained in light increases as the frequency increases from infrared through red to blue and ultraviolet light. However a solar cell must throw away some of these frequencies. It can only absorb light which is above a certain energy threshold, called the *cell threshold*. Light with energy content below that threshold will simply pass through the cell or be reflected.

Ideally, then, one might want to choose the lowest-feasible threshold in order to utilise as much of the solar spectrum as possible. There is, however, another difficulty. The cell threshold determines the cell voltage. If the threshold frequency is very low, the solar cell will provide a low-output voltage. When light is absorbed with an energy higher than the threshold, all the excess energy is simply thrown away, wasted. So setting the threshold too low wastes energy too. Thus the threshold must be set at an optimum level.

Most commercial solar cells use silicon as the semiconducting layer. Silicon does not represent the optimum solar absorber but it is relatively easy to work with, it is extremely abundant and therefore cheap, and it benefits from an accumulation of experience with the material as a result of its use in the manufacture of transistors and microchips.

Other materials are being introduced. Among the most promising are cadmium telluride and copper indium selenide. Gallium arsenide has also been used in space. Silicon, however, appears likely to form the backbone of the solar cell industry for the immediate future.

Types of solar cell

Microchips and transistors are universally fabricated using slices cut from perfect crystals of silicon. These crystals are carefully grown under controlled conditions and are expensive to produce.

Solar cells can be made with single crystal silicon too. Indeed, cells using this material have provided the best performance of any silicon solar cells, with solar-to-electrical conversion efficiencies of up to 24%. Long-term performance of single crystal silicon cells is good too, but the cost of the single crystal material remains relatively high.

In an attempt to bring fabrication costs down, many manufacturers have experimented with alternatives to single crystal silicon. One of the most widely used is a form called *polycrystalline silicon*; this is silicon made up of lots of tiny individual crystals instead of one large crystal. Such material is much cheaper to produce but has proved less efficient than the single crystal material. However it does produce reliable and stable cells, at a significantly lower cost. Efficiencies of over 18% have been achieved.

Cheaper still is a completely non-crystalline form of silicon called *amorphous silicon*. This was initially found to suffer from a serious degradation problem when exposed to light, an effect which reduced efficiencies by 20–40%. With extensive redesign, it has now proved possible to circumvent the most serious aspect of this problem. The amorphous cell does still suffer degradation of around 20% but its output eventually stabilises. Cell efficiencies of around 13% can now be achieved (after degradation has taken place).

The cheaper and simpler amorphous silicon fabrication process has allowed some more complex solar cell designs to be developed. For example, some amorphous cells have been fabricated as three cells one on top of the other, designed to absorb first blue, then green and finally red light. This three cell design offers the potential for higher efficiency than a single cell absorbing the whole spectrum.

All silicon solar cells require extremely pure silicon. The manufacture of pure silicon is both expensive and energy intensive. The traditional method of production requires 90 kWh of electricity for each kilogram of silicon. Newer methods have been able to reduce this to 15 kWh/kg. This still means that a silicon solar cell takes 2 years to generate the energy needed to make it.[5] This compares with around 5 months for a solar thermal power plant.[6]

Silicon-based solar cells dominate the market today. Single crystal and polycrystalline silicon cells remain the most popular, accounting together

Table 13.1 *Solar cell production, 2003, by region*

	Production (MW)
Japan	364
Europe	193
USA	103
Rest of the World	84
Total	744

Source: Renewable Energy World.[15]

for 89% of production in 2003. Of that, polycrystalline cells accounted for 62%.[7] Amorphous silicon adds a further 3%. These materials are likely to remain dominant for several years. There are alternatives to silicon. Most promising of these are cells fabricated from cadmium telluride or from copper indium selenide. Manufacture of solar cells based on these materials has begun, with a total production of 7 MW in 2003.[8]

Fifty years of experience with silicon solar cells means that long-term performance can now be assessed with accuracy. Modern silicon solar cells sold in panel form for installation on roofs come with a 25-year warranty. Longer lifetimes still should eventually be plausible. The long-term performance of newer materials has yet to be established.

Solar cell manufacture

Solar cell production is a highly technical process and this has severely limited the number of companies involved in the industry. Virtually all global solar cell production in 2003 was carried out by just ten companies. The largest, by far, of these was the Japanese company Sharp.

Regionally, Japan has come to dominate production, followed by Europe, the USA and the rest of the world. Production figures for 2003 are broken down regionally in Table 13.1. Total global production was 744 MW. That was 32% higher than in 2002. This rate of growth in production has been typical over the past decade.

Solar panels and inverters

A single modern silicon solar cell will produce between 2 and 3 W of power depending on its size. This will equate to between 3 and 5 A at 0.6 V. In order to provide a usable current and voltage, groups of cells are connected both in series and in parallel. For example, 36 solar cells connected in series will provide an output of about 20 V, suitable for a battery charger designed to recharge a 12 V battery.

For grid-connected applications, more cells are necessary. Typical units are designed to provide up to 200 W. The individual cells are normally mounted behind a glass protective barrier similar to a vehicle windscreen. The whole assembly is then encapsulated to protect it from the weather and framed with aluminium extrusions. Such assemblies are called *solar panels*.

A solar panel provides a stable direct current output. If this is to form a part of a grid-connected solar power system, perhaps on the roof of a household, it must be converted to AC at the grid voltage. This is carried out by an inverter. A typical household system will require a 2 kW inverter.

Solar cell deployment

There are three key ways of deploying solar cells. These are referred to as residential photovoltaics, utility array photovoltaics and solar concentrators. In the early years of the twenty-first century residential photovoltaics have become the most important but all three can, and may have a part to play in the future of solar electricity generation.

Utility photovoltaic arrays

A utility photovoltaic array is solar cell-based power plant with a generating capacity similar to that of a fossil fuel power plant. Construction of such a plant would involve an enormous number of individual solar cells, mounted in solar panels, and the solar panels themselves mounted in groups, each group having its own support structure.

The groups of cells would probably be fitted with a system to track the sun across the sky. Both single- and double-axis tracking are possible, the latter providing the best solar input, but at the expense of greater complexity.

Costs remain too high for this to be a competitive means of generating electricity but demonstration projects in the USA have shown that it is theoretically feasible. Initial commercial plants might have a capacity of 20 MW.

The potential for this type of technology is vast. In the USA, price permitting, it could provide up to 10% of grid-connected utility generation, or up to 200 GW.[9] Although this represents a massive market, it is in the developing world that they may find their greatest application. The small, modular nature of the utility array makes it ideally suited to remote regions where grid connection is either too expensive or geographically impossible.

Solar concentrators

A solar concentrator uses a lens or reflector to capture sunlight from a wide area and focus it onto a small area where a photovoltaic convertor device is located. Sunlight concentration may be as little as 2× or as high as 2400×. Like utility-scale photovoltaic arrays, solar concentrators are essentially a large-capacity deployment technique.

Concentrators require much smaller quantities of semiconductor material than conventional photovoltaic arrays. As a consequence it becomes cost effective to use the most efficient material available, even if this is much more expensive than the material used in large photovoltaic arrays. With only a small area of semiconductor, most of the concentrator is made from relatively cheap and readily available materials. This means that scaling up to larger sizes is easy and economical to achieve.

There are various ways of building concentrators. These range from arrays of small cells, each with a lens focusing sunlight from a small area onto a tiny photovoltaic device, to a 10–20 m parabolic reflector collecting sunlight and concentrating it at a small central receiver where the photovoltaic convertor is situated. Common to them all is the need to track the sun in order to achieve good performance, because concentrators generally rely on the incident sunlight being perpendicular to the actual solar cell.

While concentrators have yet to gain much of a foothold terrestrially, they are attracting the attention of the space industry. Their advantage in space over conventional photovoltaic arrays is a reduced exposure to radiation damage because the sensitive photovoltaic convertor is shielded inside the device. This gives them, potentially, a much longer life.

Residential photovoltaic arrays

The third category of photovoltaic applications, residential photovoltaic arrays, forms the most significant area of photovoltaic expansion as a result of major national programmes, particularly in Japan, Germany and the USA, to install solar cells on the roofs of homes.

The residential array is a type of distributed generation. As such it offers some important benefits. The power from the rooftop array is delivered directly to the place where it is needed. This is the most efficient way of generating power because there are no transmission and distribution losses involved. Further, extensive local deployment of this type reduces the need to increase transmission system capacities as demand grows.

Rooftop deployment avoids the problem of finding space to install solar arrays. The space, the roof, is already available. In addition the roof structure provides the support for the array so the expense of a dedicated support structure is avoided.

Placing solar arrays on the roofs of existing buildings will generally be a compromise because the buildings will rarely offer the optimum orientation and inclination for solar collection. Even so efficient production is possible. Far better is to design modern buildings with solar generation in mind. The solar arrays can then become the roofs, saving in building materials. They may exploit appropriately oriented wall surfaces too.

As the solar panel will only generate power during daylight hours, grid connection is essential to provide supply at nighttime. For this reason building arrays, although designated residential arrays, are actually most effective on commercial buildings where the daily cycle matches closely that of the sun.

Single household arrays offer the simplest application but the most cost effective may be a group deployment where tens or hundreds of units are fitted in a compact area. Such projects could be underwritten or owned by a local utility or distribution company. Equally they could be owned and operated by a residents' co-operative. The larger deployment attracts economies of scale that are not available for individual householders.

Environmental considerations

Solar power is considered to be one of the most environmentally benign methods of generating electricity. Neither solar thermal nor solar photovoltaic power plants generate any atmospheric emissions during operation. A photovoltaic installation makes no noise either, and a solar thermal plant very little. Nevertheless both types of plant do have an environmental impact.

On a utility scale, both types of solar power plant require a significant amount of space, more than that required by a fossil fuel power plant. However the best sites for such plants are likely to be in arid areas where this should not pose a problem. Construction of a large plant is likely to involve some local environmental disruption. Once in operation there may be some benefits locally from the shade created by the arrays of solar collectors.

When solar panels are installed on rooftops or incorporated into new buildings they share space used for other purposes. Retrofit of solar panels can be unsightly but where a building has been designed to incorporate solar panels there is no excuse for any negative visual impact.

This type of deployment has environmental benefits because it reduces the need to central power station capacity, it reduces the need to reinforce transmission and distribution systems, and it provides electricity at the point of use so energy losses should be much lower than when power is transmitted many kilometres.

Solar thermal power plants rely on conventional mechanical and electrical components. There may be spillages of heat transfer fluid but these should be easy to control. Otherwise their construction, operation and decommissioning should be easily managed without major affects on the local environment.

Solar photovoltaic devices use less commonplace materials. The predominant material for solar cells today is silicon. This is very energy intensive to produce in its pure form. Lifetime analysis of photovoltaic systems show a relatively high level of emissions of carbon dioxide and other atmospheric emissions as a result of the emissions from the predominantly fossil-fuel-fired power plants generating the electricity used in the production of the silicon.

Lifetime analysis of photovoltaic generation suggests that such a plant will release between 100 and 170 g of carbon dioxide for every kilowatt-hour of electricity it generates. This is much higher than from a solar thermal power plant for which the equivalent figures are 30–40 g/kWh. It is, nevertheless, much lower than a gas-fired power station (430 g/kWh) or a coal-fired power station (960 g/kWh).[10] In the future this impact should be reduced as global renewable capacity grows and with it a wider availability of cleaner electricity.

The large-scale deployment of solar cells will involve much larger quantities of semiconducting material than has been manufactured for microprocessors. Some newer semiconductor materials contain toxic elements; the cadmium in cadmium telluride is a good example. This semiconductor is a stable material but it will be important to ensure that conditions cannot occur which would permit cadmium to enter the environment. This will be particularly important when a plant is decommissioned. The processes involved in the manufacture of both silicon and other solar cells involves toxic organic chemicals and these, too, have to be strictly contained.

Financial risks

The risks associated with the deployment of solar photovoltaic and solar thermal power generation technologies are primarily the risks always associated with new technologies. These relate to reliability, plant lifetime, and long-term operation and maintenance costs.

Solar photovoltaic devices are now well understood and the reliability of the predominant silicon technology has been broadly established. Solar thermal plant experience is limited to nine plants in California and while the data from these plants is encouraging it cannot be considered exhaustive, particularly since any new plants are likely to use modified plant designs to reduce costs.

The Californian plants all use solar trough technology. The two other solar thermal technologies, solar towers and solar dishes, have yet to be tested commercially so any gauge of performance must rely on data from demonstration projects. The perceived risk of these technologies is likely to be higher, but all three solar thermal technologies can expect to be viewed with a degree of scepticism by financial institutions.

From a resource perspective, solar energy is well understood. Solar insolation records exist just about everywhere so there should be no problem establishing the expected solar input at any site on the earth.

Even then, unforeseen effects can occur. The solar plants at Kramer Junction in California recorded drops in gross solar output during 1991 and 1992 that were attributed to the global ramifications resulting from the eruption of Mount Pinatubo in the Philippines.

There could also be a risk associated with the diurnal nature of solar power generation. Solar power can replace conventional sources of power during daylight hours. But without some form of energy storage it cannot supply power when it is dark.

It can, therefore, be argued that solar power is best deployed for peak power generation (demand peaks in hot countries often coincide with highest temperatures where there is widespread use of air conditioning). There is a danger, if it is deployed in this way, that future demand management programmes may reduce the peak demand level. Then the marginal value of the solar output may fall. The significance of this will depend on the mode of operation of the solar project and the details of any power purchase agreement.

The cost of solar power

Solar thermal and solar photovoltaic power plants share a number of features such as short deployment times and additional benefits from dispersed deployment that affect the cost and value of both technologies. However the technologies themselves have different roots and the costs associated with them have to be considered separately.

Solar thermal costs

Table 13.2 lists costs for solar thermal power plants estimated by the Sandia National Laboratory and the National Renewable Energy Laboratory, both run under the auspices of the US Department of Energy.

Table 13.2 *Solar thermal costs*

	Capital cost ($/kW)	*O & M costs($/kW)*	*Levelised energy Cost ($/kWh)*	
			2000	*2010*
Solar trough	2900	1.0	0.11	0.09
Solar tower	2400–2900	0.7	0.09	0.05
Solar dish	2900	2.0	0.13	0.06

Note: The levelised energy cost is for private financing.
Source: US Department of Energy.

The only one of the three technologies listed that is operating in a commercial environment is the solar trough technology, exemplified by the nine plants built during the late 1980s and early 1990s in California. Operational costs have fallen at these plants in recent years and performance has increased so they may well be generating electricity at around $0.11–$0.12/kWh, in line with predictions in Table 13.2. This is too expensive for the technology to compete in the bulk power market in the USA but is low enough to enable it to compete in niche markets. Perhaps more importantly, it can also compete in the peak power market, which is where power from the Californian plants is sold.

The other two solar thermal technologies are currently at an earlier stage of development than solar trough technology. Consequently they are not operating under commercial conditions.

Costs for all three technologies are expected to fall. By 2010, the Department of Energy (DOE) predicts that solar towers will be capable of generating power at a levelised cost of $0.05/kWh, solar dishes at $0.06/kWh and solar troughs at $0.09/kWh. These levelised prices have been estimated on the basis of projects being built by an independent power producer with private financing. Other estimates have predicted an even lower cost for solar trough power plants, perhaps as low as $0.06/kWh by the middle of the next decade.

Costs are likely to be lower still for an ISCC power plant. A World Bank assessment put the cost of a near-term ISCC plant based on solar trough technology at 1080/kW, and the generating cost at less than 0.07/kWh[11]. This could fall to 0.05/kWh over the longer term.

Plants such as solar trough facilities with large solar arrays could be cheaper to build in the developing world where labour costs are lower than in the developed. A 100-MW solar trough plant could cost 19% less in Brazil than in the USA, for example.[12]

Solar photovoltaic costs

The main market for solar photovoltaic technology in 2003 was grid-connected residential and domestic installations. These accounted for 365 MW of total annual production of 744 MW, or roughly 50%.[13]

The cost of a grid-connected solar photovoltaic system based on silicon can be divided roughly into thirds. One-third is for the actual silicon to make the cell (the module), a further one-third for the manufacture of the solar cell and panel or module, and one-third for installation and ancillary equipment.[14]

In the USA in 2003, the cost of an installed rooftop system of this type was $6500–$8000/kW (see Table 13.3). This compares with $7000–$9000/kW in 2001 and $12000/kW in 1993. Even so, this makes solar photovoltaic technology one of the most expensive available today for generating electricity.

Table 13.3 *Solar photovoltaic costs*

	Photovoltaic module ($/kW)	*Installed AC system ($/kW)*
1993	4250	12,000
1995	3750	11,000–12,000
1997	4150	10,000–12,000
1999	3500	9000–11,000
2001	3500	7000–9000
2003	3000	6000–8000

Source: Renewable Energy World.[16]

The cost of the solar cell accounts for a major part of the overall cost. Table 13.3 shows, this is between one-third and one-half of the total cost. Newer technologies may offer hope of reduced costs. Amorphous silicon and cadmium telluride modules were selling for $2000/kW and $3000/kW in 2003. The manufacture of silicon designed specially for solar cell applications may also reduce costs of silicon further.

Apart from the introduction of new technologies, the main hope for a reduction in the cost of solar cells comes from economies of scale. This effect is already bringing prices down slowly and global capacity rises. Government sponsored schemes to encourage the use of grid-connected photovoltaic arrays in commercial and domestic situations in countries like Japan, Germany and the USA are helping to increase demand.

The cost of electricity from solar photovoltaic power plants remains high. At an installed cost of $5000/kW, electricity probably costs around $0.25/kWh. This can be competitive with the peak power costs in somewhere like California but is way above the cost of base-load power, $0.025–0.050/kWh, in markets with well-developed infrastructures. Nevertheless the cost has reduced to a point where widespread installation is feasible.

End notes

1 In fact the Californian plants have achieved 20% conversion efficiency.
2 This is a World Bank estimate.
3 Photovoltaics come down to earth, Bill Yerkes, Modern Power Systems (July 2004) p. 30.
4 Renewable Energy Technology Characterization, The US Department of Energy and the Electric Power Research Institute (Topical Report 109496) 1997, available at http://www.eere.energy.gov/power/pdfs/techchar.pdf

5 Refer *supra* note 3.
6 Solar Thermal Power 2020, Greenpeace, 2004.
7 PV Market Update, Paul Maycock, Renewable Energy World (July–August 2004).
8 Refer *supra* note 7.
9 Refer *supra* note 7.
10 US Energy Information Administration estimate.
11 Lifecycle figures are taken from Benign Energy? The Environmental Impact of Renewables, published by the International Energy Agency.
12 Cost Reduction Study for Solar Thermal Power Plants, Final Report, World Bank (May 1999).
13 Refer *supra* note 6.
14 Refer *supra* note 7.
15 Refer *supra* note 3.
16 Refer *supra* note 7.

14 Ocean power

This chapter covers three very different technologies, all designed to extract energy from the world's seas and oceans. The first, ocean thermal energy conversion (OTEC) uses the difference in sea temperature between the surface and the deep ocean to drive a heat engine and generate electricity. The second, wave energy conversion, extract energy from ocean waves, while the third, called *ocean current* (it is also called *ocean stream* and *tidal stream*) conversion uses devices that operate like undersea wind turbines extracting energy from moving currents of water.

All these technologies are in development and while each has been demonstrated at a pilot scale, none is yet ready for commercial deployment. Wave energy conversion is the oldest of the three. The first patent for a wave energy device was issued in France at the end of the eighteenth century. However the modern history of wave energy conversion dates from the 1970s. OTEC was first suggested in the 1880s, again by a Frenchman, and has been the subject of intermittent development since the 1930s. Ocean current technology is the most recent, with the bulk of its development taking place during the last decade. All are renewable technologies which hope to gain support from twenty-first century sustainable energy programmes.

Ocean energy resource

These three technologies harness energy that is available in three different forms in seas and oceans. OTEC takes advantage of what is essentially a secondary source of solar energy. It relies on the fact that tropical and subtropical seas absorb considerable quantities of energy from the sun, energy which elevates the surface water temperature. Deep water does not receive any solar radiation and so remains much cooler. This temperature difference can be used to drive a heat engine. A temperature difference of at least 20°C is normally required to make OTEC effective. This can usually be achieved provided there is a depth of water greater than 1000 m available.

The world's oceans absorb around 4000 times the amount of energy consumed today by human activity.[1] OTEC plants can probably only achieve 3% efficiency but even at that rate they could theoretically provide 13 times world energy consumption. The greatest resource is located in tropical and subtropical regions and much of the available energy will be far out to sea where it cannot easily be transported back to land. Promising sites exist in a number of places, particularly close to island communities. Schemes have also been devised that would take advantage of energy available far from land to produce hydrogen which could then be transported back for use.

Wave energy converters extract energy from ocean waves. These are surface waves generated by the passage of wind across water. Since global winds are caused by temperature and pressure differences in the atmosphere resulting from solar heating, wave energy is essentially another, in this case tertiary, form of solar energy.

The energy carried by a wave depends on the wind speed and the distance a wave travels in the wind. The swifter the wind and the longer the wave travels with the wind at its back, the greater the energy the water absorbs. The most energetic waves are found in the world's great oceans where the wind can blow uninterrupted for thousands of kilometres. Once formed, waves continue to travel with little energy loss over deep oceans until they reach land. The western coasts of the Americas, Europe and Australia, New Zealand and some Indonesian islands have particularly good wave regimes. Waves will typically deliver 10 kW–50 kW for each metre of crest length.

Once waves enter shallower water they increase in size. The size and direction is also affected by coastal features and waves can become focussed by natural or artificial features. This can be useful where energy is to be extracted.

Total global wave energy potential has been estimated at 1–10 TWh.[2] The best wave energy conditions are found between 40 and 60 degrees latitude, north and south of the equator where the wind regimes are strongest. Here wave energy is usually over 30 kW/m and may reach 100 kW/m. The region 30 degrees either side of the equator provides wave regimes with lower energy content but the resource in this part of the world can be much steadier than that farther north or south. A European Union (EU) assessment of its regional wave energy resource estimated that 320 GW of capacity was available.

Ocean currents are generated in a variety of ways. Many are the result of tidal motion, a consequence of the varying gravitational attraction on the earth of the sun and the moon. Others, such as the Gulf stream, result from temperature differences or differences in salinity.

Currents, particularly when tidal in origin, are affected by the shape of a coastline and the contours of the seabed. Strong currents are often found between islands and the mainland and at the entrances to lakes and lochs. Many good ocean current sites are available close to land. These can be harnessed relatively easily to provide energy.

The total power contained in ocean currents is between 2 TW and 5 TW[3] but only a fraction of this is accessible. EU estimates suggest that the coastal waters around EU countries could provide 34–46 TWh/year of electricity from near-shore sites and 120–190 TWh/year from offshore sites.[4] The global generation capacity is around 2000 TWh.

Ocean thermal energy conversion

OTEC relies on the principle exploited in most forms of electricity generation that a source of heat and a source of cold[5] can be used to drive an engine. In the case of OTEC the source of heat is the surface of a tropical or subtropical sea while the source of cold is the deep sea.

The possibility of extracting energy from the sea in this way was recognised in the latter half of the nineteenth century and the first practical system was proposed by the French inventor D'Arsonval in 1881. D'Arsonval's system employed a closed cycle ammonia turbine and it was finally demonstrated in 1979 in a small pilot project of Hawaii. However a different, open cycle system was tested during the 1930s by another Frenchman, Georges Claude. Claude's system proved the theory on which it was based but was not successful commercially. Work was not revived on his system until late in the twentieth century.

Tropical oceans and seas have surface water temperatures of between 24°C and 33°C. Below 500 m, the temperature will drop to between 9°C and 5°C. This provides a maximum exploitable temperature difference of 28°C. In practice the temperature difference is likely to be closer to 20°C, providing a theoretical energy conversion efficiency of 6.7%.[6] When account is taken of the need to pump cold water up from the depths, efficiency falls to 2–3%.

Though cool water may be available at 500 m, in practice the depth of 1000 m is normally considered necessary. If this is to be made accessible from land, a very long cold water pipe will be required to pump the deep water to the plant. This pipe will need to be 2000 m, or more, in length and if care is not taken, the cold water will become warmed before it reaches the plant. The alternative is to build an OTEC plant on a floating platform from which the cold water pipe stretches vertically downwards. Even with this arrangement, the cold water pipe will need to be 1000 m long.

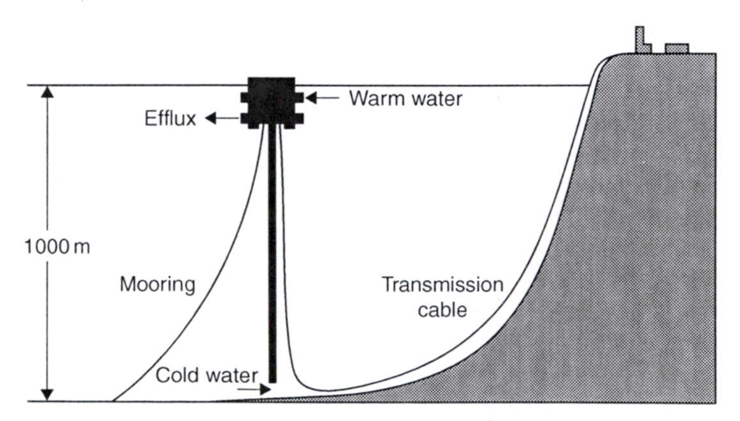

Figure 14.1 *Schematic diagram of a floating OTEC plant*

In order to generate 1 MW of electricity, an OTEC plant requires 4 m^3/s of warm seawater and 2 m^3/s of cold seawater. This will require a cold water pipe of around 11 m in diameter to supply a 100 MW plant, the largest size considered practical.[7] The discharge of mixed hot and cold water from a plant of this size would be equivalent to that of the Colorado River in the USA discharging into the Pacific Ocean. Such massive quantities of water could have significant environmental impact.

Open and closed cycle ocean thermal energy conversion

The siting of an OTEC plant, either onshore or offshore, represents one of the key decision for any proposed project. The other key decision is the type of cycle to use. There are two principle options, an open cycle plant or a closed cycle plant. Hybrids of the two have also been proposed.

A closed cycle OTEC plant employs a thermodynamic fluid such as ammonia or a refrigerant like freon. This is contained in a completely closed system including the plant turbine. Hot surface seawater is used to evaporate the fluid and the vapour is then exploited to drive the turbine. The vapour from the turbine exhaust is condensed using the cold, deep ocean water, and returned to the beginning of the cycle where it can be reheated. A 50 kW closed cycle OTEC plant was built in Hawaii in 1979 and operated for a few months. A consortium of Japanese companies has also operated a 100 kW closed cycle OTEC plant in Nauru. Again this plant operated for only a few months to prove the concept. Neither was large enough to be commercially viable. Indeed, closed cycle OTEC is unlikely to be commercially viable in sizes of less than 40 MW.

In an open cycle OTEC system the seawater itself is used to provide the thermodynamic fluid. Warm seawater is expanded rapidly in a partially evacuated chamber where some of it 'flashes' to steam. This steam is then used to drive a steam turbine. From the exhaust of the turbine, the vapour is condensed using cold seawater. The vapour produced by flashing warm seawater is at a relatively low pressure so it requires a very large turbine to operate effectively. Practical limitations mean that the largest open cycle turbine that can be built today is around 2.5 MW, much smaller than for a closed cycle system.

One of the major advantages of the open cycle system is that the water condensed from the turbine exhaust is fresh, not salt water, and so the plant can also serve as a source of drinking water as well as electricity. A 210 kW open cycle OTEC pilot plant operated in Hawaii between 1993 and 1998.

In a hybrid OTEC plant warm seawater is flashed to produce steam and this steam is then employed as the heat source for a closed cycle system. This system is more complex that either of the other cycles but it marries the compact closed cycle system with the ability to produce drinking water.

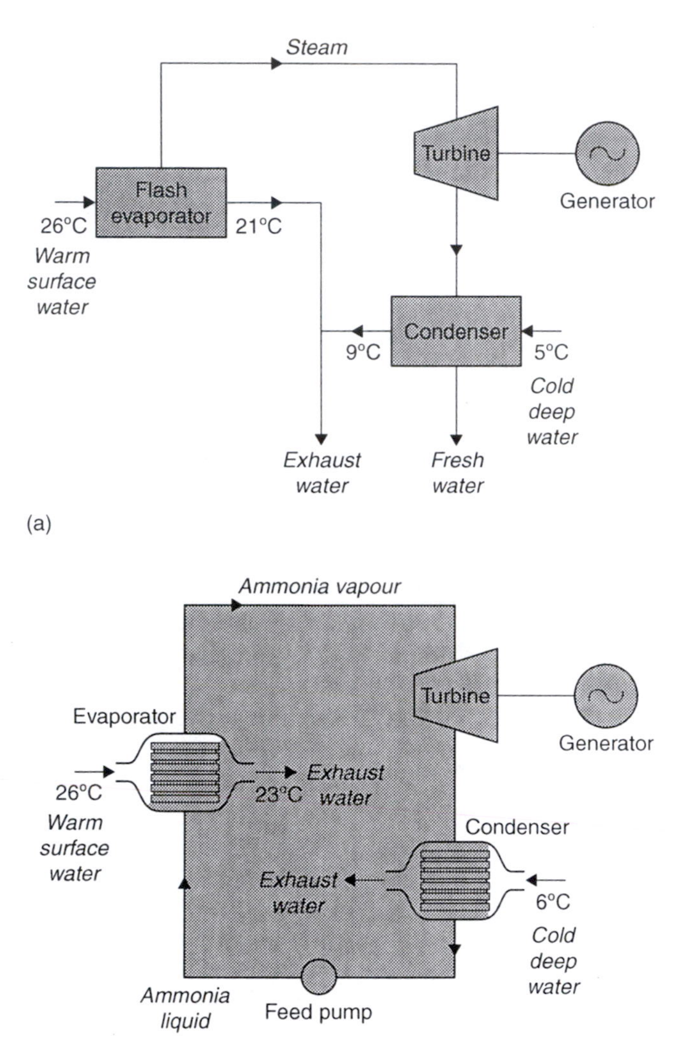

Figure 14.2 *Block diagrams of (a) an open cycle OTEC plant and (b) a closed cycle OTEC plant*

Technical challenges

The major challenge facing OTEC is the development of cold water pipe technology. The cold water pipe has to pump water from a depth of around 1000 m. For a floating OTEC plant the pipe will be at least this long. Land-based plants will require significantly longer cold water pipes. A 40 MW project proposed for Hawaii would involve a cold water pipe of over 3.5 km in length. Designing pumps and piping capable of delivering the volumes of water required over this distance without significant temperature rise represent a considerable engineering challenge.

Heat exchangers are also important components of an OTEC plant. The two heat exchangers are likely to be the largest and most costly components and require careful optimisation. As already noted, the turbine in an open cycle OTEC plant is also large. Special low-speed turbine designs will be needed to achieve high outputs from this type of OTEC plant.

Offshore OTEC plants will require deep-water moorings. Modern offshore oil and gas expertise should provide a good starting point for developing such moorings but further work will be needed to tailor mooring systems to the needs of the OTEC facility.

Hybrid applications

The OTEC plant is designed primarily to generate electricity. As such is could provide significant souce of power to a small island community where the conditions for OTEC exploitation exist. Such communities will also often benefit from fresh water production from an open cycle or hybrid OTEC facility. This combination will prove important for the future of OTEC.

There is a further resource available from an OTEC plant, a supply of nutrient-rich and bacteria-free deep ocean water. This can be used for forms of aquiculture as well as to provide cooling. When combined with electricity and drinking water production, aquiculture could make OTEC more attractive economically.

Browsing ocean thermal energy conversion

The short-term application of OTEC will be for land-based or floating inshore plants providing services to a local community, with power possibly supplied to a grid system. However most OTEC potential is far offshore. This could be exploited by browsing OTEC plants, which cruise the oceans looking for the hottest surface water temperatures. Such plants would not be able to transfer electricity directly to land. However they could generate hydrogen and potable water, both of which could be stored for later transportation to land. Current costs would not make this economically practical today.

The environmental impact of ocean thermal energy conversion

The main environmental impact of an OTEC plant results from the pumping of water from below 1000 m and then returning it at a much lower depth. The surface water for an OTEC plant is taken at around 20 m depth, so the mixed hot and cold water must be returned at around 60 m depth to prevent it returning directly to the hot water input.

The volumes of water involved are enormous and the movement of water from lower to upper regions of the seas and oceans could have a significant impact on the local marine environment. There is little evidence available today to indicate what the effect would be, but the danger of this impact is likely to limit the exploitation of OTEC, at least until extensive environmental studies have been carried out.

The warm water heat exchanger in an closed cycle OTEC plant is likely to be subject to biofouling. This must be prevented to maintain efficiency. The only way to prevent biofouling is by chemical treatment, probably with chlorine, leading to some release into the sea. Such releases would need to be closely monitored and would have to fall within legal limits.

Construction of an OTEC plant will lead to some seabed disruption but this is likely to be short term and relatively minor. There will also be the danger of small releases of oil and, in the case of a closed cycle OTEC plant of thermodynamic fluid. The impact of such releases should be of a similar to those from existing offshore facilities.

The cost of ocean thermal energy conversion

OTEC is still under development and any costing must be considered extremely tentative. Like many renewable technologies, OTEC is capital intensive with capital costs starting at around $4 000/kW. This is expensive for a power plant but may appear less so if drinking water production is taken into account. However pre-commercial demonstration of the technology is still required to prove that it is viable. There is too little experience to provide any realistic estimate of the cost of electricity generated from an OTEC plant.

Wave energy

All seas contain energy in the form of surface waves which can be exploited by wave energy conversion devices. Not all seas provide an economically exploitable resource. A good wave regime will normally be the first consideration. Such regimes are found principally on western coasts facing the world's great oceans. Wave energy exploitation is limited to coastal and near-shore sites, so the opportunity for deployment is restricted. Even so, it is possible that wave energy could supply up to 10% of global electricity demand.[8]

The development of modern wave energy conversion technology started after the oil price rises of the 1970s. Much work was carried out in the UK but national funding was withdrawn in 1989. Work continued in other countries in Europe and in the USA and Japan. The environmental concerns

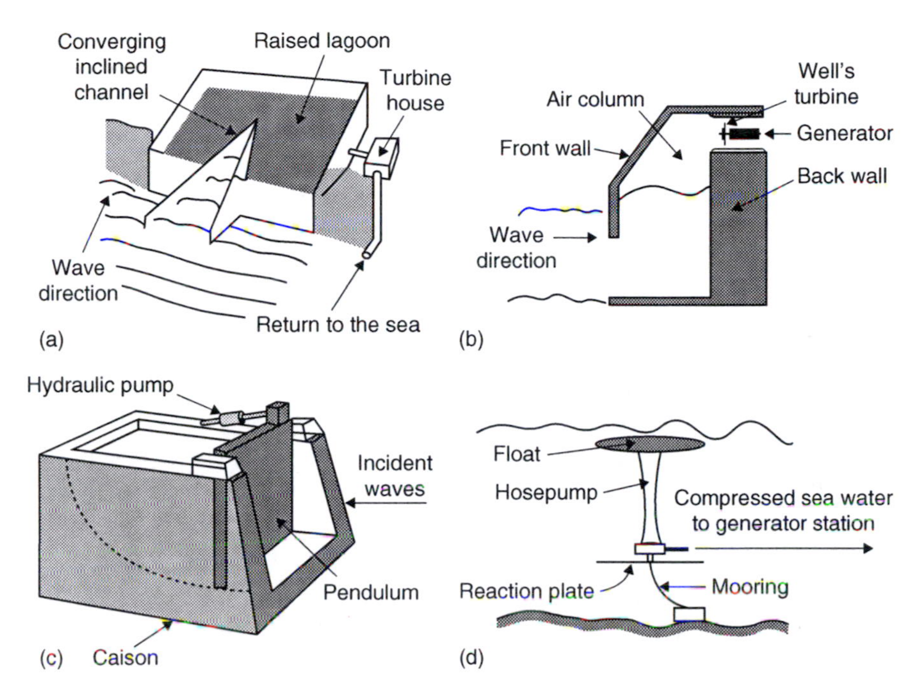

Figure 14.3 *Wave energy conversion devices. (a) Tarpered channel device (Tapchan); (b) Oscillating water column (OWN); (c) Pivoting flap device (the pendulor); (d) Heaving buoy device (the hosepump)*

of the 1990s added further impetus to the wave energy industry and the UK resumed funding in 1999. By the beginning of the twenty-first century a wide variety of different wave generation devices were under development.

Unlike virtually all other power generation technologies, wave energy conversion requires completely unique energy conversion devices. These devices have to convert the wave motion at the sea surface into electricity. Engineers who have addressed the problem have devised a range of novel solutions. These can be categorised in a number of ways but perhaps the simplest is to divide them in to two groups. The first comprises shore and bottom mounted near-shore devices and the second comprises offshore devices.

Shore and near-shore wave converters

1. Oscillating water columns

Perhaps the most widely tested of shore and near-shore devices is the oscillating water column. If a tube, sealed at one end, is placed so that its open end is just beneath the surface of the sea, as waves pass the tube, the level of water inside the tube will rise and fall, alternately compressing and expanding the air column within the tube.

If, instead of a seal, the upper end of this tube is open and houses a device that acts like a wind turbine, then the moving water will cause the air to move in the tube and this will make the turbine rotor turn backwards and forwards. This air movement forms the basis for an oscillating water column wave energy converter.

Oscillating water columns can either be shore or bottom mounted. They normally comprise some form of concrete structure, which is designed to create an enclosure containing air, which is open to the sea at the bottom. A special type of turbine called a *wells turbine* is frequently mounted at the top. This can derive continuous power from movement of air both up and down without the need for a complex arrangement of valves.

Oscillating water columns have been tested in many parts of the world including Europe and Japan. The wells turbine, developed in the 1970s, has most usually been employed in these prototypes but newer bi-directional turbines with greater efficiency are under development. The economics of oscillating water column converters and of other shoreline devices can be improved if they are built into breakwaters.

2. Tapered channels

Another approach to wave energy conversion uses and amplifies the height of a wave in order to create a head of water which can be used to drive a conventional low-head hydro turbine. Devices like this usually employ a tapered channel with its mouth open to the sea. The side walls of the channel rise above the normal sea level and beyond them is a reservoir.

Waves travelling towards the coast are focussed into the channel. As these waves flow along the channel they become more and more restricted by the taper and this forces the height of the wave to increase, until water starts to fall over the upper edges of the channel walls. This water is captured, creating a reservoir of water, which is above the sea level. This water can then be run back into the sea through a hydro turbine, generating electricity.

A system of this type, called *tapchan*, was built on the Norwegian coast in the late 1980s. The technology is relatively simple but construction costs can be high. Deployment is restricted by the need for a relatively low tidal range. Otherwise the operation of the converter is compromised.

3. Oscillating flaps

The energy contained in moving waves is sufficient to cause a pendulum or flap to move backwards and forwards, and this too has been used as the basis for a shore-based wave energy convertor. The best known converter of this type is a Japanese device called the *pendulor* which comprises a box, open to the sea on one side, but with the open side closed using a flap hinged horizontally from the top. When waves strike the flap they cause it to oscillate to and fro like a pendulum and this motion can be converted into electricity using hydraulic[9] rams. Small devices of this type have been built and tested and a plant of around 200 kW has been designed for a site in Sri Lanka.

Offshore devices

The three types of device discussed above can all be exploited offshore provided they can be moored so that they remain stationary relative to the waves. However most offshore devices try to exploit the wave motion in different ways.

1. Float pumps

The hosepump, developed in Sweden, is based on an elastic tube that changes its internal volume as it is stretched. One end of the tube is sealed and attached to a float while the other end is open and, is connected to a moored plate close to the bottom of the sea. As waves pass the device, the float moves up and down, alternately stretching and relaxing the tube. This pumps water in and out of the lower end of the tube and the pressurised water is used in a hydraulic energy conversion system to generate electricity.

A wave power float pump developed in Denmark takes a slightly different approach. In this case a float at the surface is attached to a rod, which bears in turn on a shaft (like a crankshaft), attached to a piston-pump device. Movement of the float up and down causes the rod to rotate the (crank) shaft, turning the vertical motion into rotary motion from which electricity can be extracted, exactly as in a piston engine.

The Archimedes Wave Swing, developed in the Netherlands, adopts a similar principle but the movement up and down of a buoyant floater is converted into electricity by means of a linear generator.

2. Ducks, wave pumps and other water snakes

There are a number of wave energy devices designed in hinged sections which all float. As waves pass these devices, the different sections move relative to one another and this differential motion is used to derive hydraulic energy, which is then converted into electricity.

The first of these is called *Salter's Duck* after British designer Stephen Salter. The prototype appeared in the 1970s but the concept is still under development. The duck has a beak-shaped float, which is fixed by a hinge to a second anchored section. The beak moves in the waves relative to the anchored spine and this relative motion is used to extract energy.

The McCabe wave pump comprises three rectangular pontoons connected through hinges. The central pontoon has a damper plate attached to it, which slows its vertical motion relative to the two outer pontoons. This generates relative motion between the three sections, which again can be exploited to generate electricity.

The Pelamis looks like, and is named after a sea snake. It comprises a series of buoyant cylindrical sections joined end to end. The device is tethered by one end. As waves pass along it, the sections move relative to one another and hydraulic cylinders extract energy from this motion. A 750 kW prototype is under construction.

3. Piezoelectric devices

A US programme is developing a wave energy converter, which is based on a piezoelectric material in sheet form. The device, called the *eel*, can produce an electric current when bent by waves. It is still in a very early stage of development.

The environmental implications of wave energy converters

Wave energy converters remove energy from waves. This means that both near-shore bottom mounted devices and offshore floating devices will calm the sea. This will be broadly beneficial since it will protect the coast from waves.

Land and near-shore bottom mounted wave converters will cause some disruption to the marine environment during construction but this should be short lived. Once in place they should have little impact. There may be a visual impact and oscillating water column converters may generate noise from their air turbines.

Floating offshore devices should be less disruptive since they will normally be built onshore and then towed to the site where they are deployed. However floating devices are likely to be a hazard to shipping and sites will need to be selected that cause minimum disruption. It may well make sense to deploy such devices at offshore wind sites; both need to be away from shipping lanes and well marked to prevent the danger of collision. A floating device could become a serious hazard if it slipped its mooring, so strict monitoring would be essential.

The cost of wave energy conversion

Wave energy conversion is still in an early development stage and it is impossible to gain a realistic idea of costs. However UK estimates have suggested that wave power may be able to produce electricity for between €0.06/kWh and €0.12/kWh. This may make wave power more economical for remote coastal and island communities that currently rely on diesel generation. Capital costs range from €800/kW to €40 000/kW though the former must be considered optimistic at this stage while the latter refers to a technology in an early stage of development. Commercial wave power plants are unlikely to be deployed before 2010.

Ocean current generation

The movement of water within oceans and seas is the basis for ocean current energy conversion. This movement may be caused by tidal ebb and flow. In this case the water movement will follow a sinusoidal variation in speed and direction, the latter normally reversing twice every 24 h.

Other currents are caused by thermal gradients. The most prominent of these is the Gulf stream which moves around 80 million m^3 of water each second[9] but there are other, lesser currents in many parts of the world. These currents usually flow in one direction only, and are relatively constant in strength.

The conversion of ocean currents into electricity involves similar considerations and technology to that employed by wind power plants. The main difference is that the energy density contained in a current of water is much higher than that of air. As a result turbines can be much smaller. For a tidal current which varies regularly a current of around 1.5 m/s is considered sufficient to exploit. Where the current is continuous in a single direction, a flow speed of 1 m/s is exploitable. The latter also offers a higher-capacity factor, around 80% whereas a tidal current will provide a capacity factor of 40–50%.

Sea currents hold one further advantage over wind; they are predictable. Thus whereas a wind farm cannot guarantee its output, the output from an ocean current power plant should be entirely predictable. This has significant implications for network operation and dispatching since a reliable source of electricity is much more valuable than an unpredictable source.

As with wind turbines, there are two different configurations of tidal stream turbine, a horizontal axis and a vertical axis turbine. Both types are under development.

Horizontal axis turbines

The horizontal axis turbine, or propeller turbine, comprises a propeller with two or more blades. The turbine can either be mounted on a tower fixed to the seabed or it can be deployed below a floating support. The former method is most suitable for shallow waters whereas that floating support can be deployed in deeper water. In order to increase the efficiency of a horizontal axis system, water flow around the turbine can be controlled using a shroud.

As a result of the high-energy density, water turbines are much smaller than wind turbines. A unit with a diameter of 10–15 m can produce between 200 kW and 700 kW. Prototypes include a 15 kW unit tested in a Scottish Loch and a 300 kW unit deployed off the coast of southern England in 2003.[10]

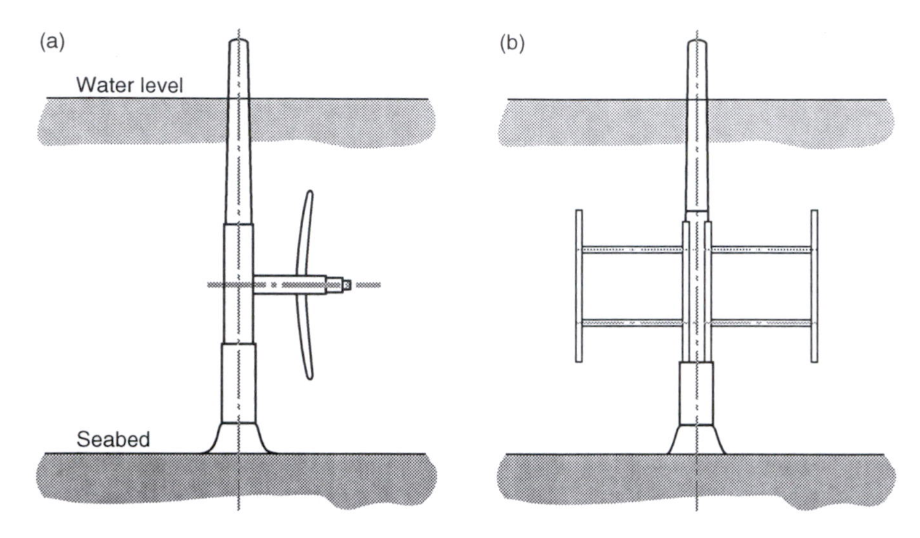

Figure 14.4 *Horizontal and vertical axis ocean current energy converters. (a) Horizontal axis turbine (axial flow) and (b) Vertical axis turbine (cross flow)*

Vertical axis turbines

The vertical axis turbine used for ocean current applications has vertical blades which are supported on struts attached to a vertical shaft. The blades are shaped so that they will rotate in a current, whatever be its direction. This is particularly useful for tidal applications where the current direction reverses regularly.

As with the horizontal axis turbine, a vertical axis machine can be either bottom mounted or fixed to a floating platform. The latter design has been tested in a 130 kW prototype in Italy.

It is possible to deploy an array of vertical axis turbines arranged like the elements of a vertical fence. When used in tidal waters this is called a *tidal fence*. The scheme allows the maximum amount of energy to be extracted from a single site. A prototype based on this concept is being planned for installation in the Philippines.

Other tidal stream energy extractors

It is possible to employ devices other than turbines to extract energy from a moving stream of water. One such device, called the *stingray*, is based on concept of the hydroplane. The stingray has a large hydroplane-like wing mounted at the end of a long, hinged arm. The angle of the hydroplane can be adjusted to control its lift and drag. If the device is placed in an ocean stream, adjusting the angle of attack of the hydroplane cyclically will force the arm to oscillate up and down, generating a hydraulic force, which can be converted into electricity. A 150 kW prototype is under development.

Ocean current environmental considerations

Ocean current energy conversion devices should normally be constructed on shore and then transported to the chosen site for installation. Seabed and marine disruption should be short and impact should be small.

More significant is the fact that an ocean current energy converter will remove energy from the ocean current, leaving it weaker. This could have a significant effect on downstream marine ecologies. A major tidal stream plant such as a tidal fence-style array of turbines would probably have a similar effect to a large tidal barrage. The effect of smaller units would be less but an environmental impact study would certainly be necessary to establish their probable extent.

The moving blades of an underwater turbine could injure or kill marine mammals and fish. Further study is required to establish how dangerous this will be. Measures similar to those needed with conventional hydropower plants are likely to be necessary in order to minimise this danger.

The other main impact of an ocean current installation will be on shipping and fisheries. Large underwater structures will form a hazard to shipping, so major shipping lanes must be avoided. Other sites may interfere with local fisheries and these too must be taken into consideration.

Cost of ocean current technology

As with the other technologies discussed in this chapter, ocean current technology is still at an early stage of development and realistic costs are difficult to establish. Some early European studies have suggested that electricity could be generated for between €0.05/kWh and €0.15/kWh. Meanwhile, a Canadian study published in 2002 concluded that Canadian technology could produce electricity for between 11 Canadian cents/kWh (for an 800 MW development) and 25 Canadian cents/kWh (for a 43 MW installation). Generation costs within these ranges would make the technology competitive with diesel generation.

End notes

1 Ocean Thermal Energy Conversion (OTEC), L.A. Vega, 1999.
2 Renewable Energy Sources: 2000–2020. Opportunities and Constraints. World Energy Council, 1993.
3 The Ocean as a Power Resource, J.D. Isaacs and R.J. Seymour, *International Journal of Environmental Studies* 4, 201–203, 1973, EU ATLAS project.

4 EU ATLAS project.
5 In thermodynamics these are usually referred to as a heat source and a heat sink.
6 Ocean Energy Conversion, M.T. Pontes and A. Falcao, Lisboa, Portugal.
7 Refer *supra* note 1.
8 World Energy Council, Survey of Energy Resources, 2001.
9 Refer *supra* note 8.
10 The company which has developed this turbine, IT Power, has suggested that up to 20% of the UK's power could be derived from tidal streams.
11 Refer *supra* note 6.
12 Green Energy Study for British Columbia, Phase 2: Mainland Tidal Current Energy (October 2002). This report was prepared by Triton Consultants Ltd for BC Hydro Engineering.

15 Biomass-based power generation

Biomass is the term used in the power generation industry to describe fuel derived directly from trees and plants. The fuel may be grown specifically for use in power generation or it may be waste such as straw from cereal farming, bagasse which is the residue from sugar cane processing, rice hulls, maize husks or wood waste from wood processing plants and forestry operations.

Biomass was the major source of energy throughout the world before the industrial revolution and it still provides 14% of global energy consumption in the twenty-first century. Most of this biomass is utilised in the developing world where its contribution to total energy consumption can rise above 40% in some countries. The majority of this fuel is burned to provide heat for domestic cooking and heating.

The use of biomass for power generation is less common. Certain types of industrial facility such as the wood, paper and sugar cane processing plants frequently utilise their wastes in either heat plants or combined heat and power plants to supply their own energy needs. There are also a small number of dedicated power generation plants which burn solely biomass. Most of these are in Europe and North America.

Figures for global biomass generating capacity are extremely difficult to establish but there was probably around 18,000 MW in 1995.[1] The capacity was projected to exceed 40,000 MW by 2010 but use is accelerating faster than this 1995 estimate suggested. At the beginning of the twenty-first century there was around 8000 MW in Europe, 7000 MW in North America and 2000 MW in the ASEAN region. The capacity elsewhere is unknown. While capacity growth remains slow, it can be expected to gather pace over the next 10–20 years because biomass is becoming recognised as a sustainable replacement for fossil fuels and in particular for coal.

There are a number of different ways of converting biomass into energy. The simplest and most widespread is to burn the fuel in a furnace and use the heat produced to generate steam which is then used to drive a steam turbine. Most existing plants of this type are extremely inefficient but newer technologies such as biomass gasification can improve efficiency significantly. It is also possible to mix a proportion of biomass fuel with coal and burn it in a coal-fired power plant, a process called *co-firing*. Liquid biomass fuels such as ethanol and organic oils can be used in internal combustion engines. Most will be used for transportation but some power generation based on stationary engines is likely.

The economics of biomass power generation depends on both the capital cost of the power plant and the fuel cost. Biomass power plants utilise the same technology as coal-fired power plants but they are more expensive because their efficiency is lower. Biomass fuel costs vary widely. Where there is a readily available source of biomass waste the cost may be low enough to make biomass competitive with coal for generating power. Dedicated energy crops are more expensive and these are generally not competitive with coal today. This situation is expected to change over the next 10–20 years. Under the special circumstances where an industrial plant burns its own biomass waste to generate power, the economics are usually highly favourable.

Simple economic comparisons between coal and biomass for power generation are misleading, however, because they ignore the environmental importance of biomass fuel. The combustion of biomass has a significantly lower impact on the environment that the combustion of coal and if this is taken into account, then the economics of biomass for power generation look much more attractive. Two factors are of particular importance. First, biomass can provide a key component in a sustainable energy future by replacing fossil fuels with a fuel which can be regenerated each year. Secondly the combustion of biomass fuel makes no net contribution to atmospheric carbon dioxide concentrations and so its use can help stabilise the level of carbon dioxide in the atmosphere. For these reasons biomass is likely to become one of the most important sources of electricity later in the twenty-first century.

Types of biomass

The global biomass resource is the vegetation on the surface of the earth. This is equivalent to around 4500 EJ (4500×10^{18} EJ) of energy. Roughly one-half to two-thirds of this is regenerated each year. At the beginning of the twenty-first century, biomass equivalent to 55 EJ was being used each year to provide energy. Estimates suggest that up to 270 EJ could eventually be utilised.[2] This could provide 50% of the global primary energy consumption by 2050.

From the perspective of power generation biomass can be divided into two categories, biomass wastes and energy crops. Biomass wastes are the most readily available forms of biomass but their quantities are limited. Energy crops, grown on dedicated plantations, are more expensive than wastes but they are capable of being produced in much larger quantities and of being produced where required. Location is important because biomass has a lower energy content than coal and cannot be transported cost effectively over great distances. It will normally be uneconomical to transport it more than 100 km. If energy plantations are established close to a biomass power plant, transportation costs can be minimised.

Biomass wastes

Biomass wastes can be divided into four categories: urban, agricultural, livestock and wood wastes. Urban biomass waste is a special category, available in relatively small quantities. It usually comprises timber waste from construction sites, some organic household refuse, and wood and other material from urban gardens. Most of this is cycled through an urban refuse collection and processing infrastructure where the biomass waste must be separated from the other refuse if it is to be burned as fuel. While separation is an expensive process there is often a fee available for disposing of the waste and this helps keep fuel costs low.

Agricultural wastes are available throughout the world and they include a number of very important biomass resources. Across Europe and North America there are enormous quantities of wheat and maize straw produced each year. These farming residues are seasonal and require storing if they are to supply a year-round fuel. Sugar cane processing produces a waste called *bagasse* at the processing plant (the waste left in the field is called *trash*) where it can easily be utilised to generate electricity. Rice produces straw in the fields and husks during processing. The shells and husks from coconuts can be used to generate electricity. Indeed wherever crops are grown and harvested there is normally some residual material which can be used as a source of energy.

There is one important caveat. From the perspective of sustainability it is important that some biomass material is returned to the soil after a crop has been harvested if the soil is to retain its fertility. If all the biomass material is removed, artificial fertilisers must then be used and this will normally prove to be unsatisfactory both environmentally and from an energy balance perspective.

Livestock residues are another special category of biomass. While there is probably the equivalent of around 20–40 EJ of livestock residue generated each year, most of this is in the form of dung which has a very low energy content and is not a cost-effective fuel for power generation. It is only where livestock is farmed intensively that it becomes economical to utilise the waste and then only when the operation is being carried out on a sufficiently large scale.

Dairy and pig farms fall into this category and it can be cost effective to use a biomass digester to convert the animal effluent into a biogas, mostly methane, which can be burned in a gas engine to generate power. A similar process occurs naturally in the landfill sites used to dispose of urban waste, and this gas can also be collected and burned. Sewage farms which treat human waste are another source of methane-rich gas.

Wood waste comprises material that can beneficially be removed from natural and managed forests to improve the health of the plantation, residues left in a forest after trees have been logged and the waste produced during the actual processing of wood in sawmills and paper manufacturing plants.

Process plant waste is the cheapest and most economical to utilise. Many sawmills and most modern paper plants burn their waste, producing heat and electricity for use in the facility. Any surplus power may be sold. Residues left after logging are generally expensive to collect and transport but they have been utilised in situations where the demand for biomass fuel is high. Similarly the removal of dead trees and undergrowth from natural forest, while improving their health and reducing the risk of fire, is an expensive process that only becomes cost effective if the value of the fuel is high.

Energy crops

Wastes, particularly agricultural and wood processing wastes, are important because they can provide a cheap source of biomass which will help biomass power generation establishes itself. Over the longer term these sources will not be able to provide sufficient fuel for a mature industry. That will have to be provided by dedicated energy crops grown on special plantations. Such plantations will also be important from an economic point of view, to provide security of fuel supply to a biomass power plant.

Energy plantations in Europe, North America and Brazil already supply the raw material for liquid biofuels such as ethanol and biodiesel. There are also experimental energy plantations in both Europe and North America which supply fuel to power or combined heat and power plants. The Philippines, too, has experimented with wood production for energy.

These experiments have already identified a number of crops that are promising energy sources. Fast-growing trees such as willow, poplar and eucalyptus are among the most encouraging. These trees can be grown in a coppicing system where the wood is harvested on a 3–7-year cycle. Grasses offer another extremely promising crop. Prairie switch grasses in North America and Miscanthus grasses in Europe can be harvested each autumn and will regrow during the following year.

The energy content of a biomass fuel depends on its water content. Most woods, when cut, contain around 50% water and will have an energy content of around 10 GJ/tonne. When the wood has been dried this will rise to 19 GJ/tonne as shown in Table 15.1. Straw from cereal crops is usually harvested with a moisture content of 15% when it had a calorific value of 15 GJ/tonne. However grasses such as Miscanthus are virtually dry when harvested in the autumn when their energy content is 19 GJ/tonne, similar to that of dry wood (see Table 15.1). Coal, by comparison, has an energy content of 27 GJ/tonne.

The economics of an energy crop depends on both the energy content of the fuel and the yield that can be obtained from each hectare of plantation. Table 15.2 shows some typical yields from energy crops obtained in the USA. As the figures indicate, switch grasses have shown yields of up to

Table 15.1 *The Calorific value of biomass fuels*

	Calorific value (GJ/tonne)
As-harvested wood	10
Dry wood	19
Straw	15
Miscanthus	19
Coal	27

Source: Energy Technology Support Unit, UK Department of Trade and Industry.

Table 15.2 *Energy crop yields in the USA*

	Yield (dry tonnes/ha/year)
Switch grass	7.7–14.3
Hybrid poplar	8.1–12.8
Willow	10.1–11.0

Source: US Department of Agriculture.

14 tonnes/ha/year whereas woody crops can yield close to 13 tonnes/ha/year. Higher yields have been recorded in both the USA and other parts of the world, as high as 27 tonnes/ha/year for woods and 20 tonnes/ha/year for Napier grass, but the US figures represent practically attainable yields under prevalent conditions today.

For the future, new strains are already beginning to show higher yields. This is a trend that is likely to continue. In fact experience is showing that it may be more cost effective to replace a coppice crop like willow or eucalyptus after it has been harvested once because new, higher-yielding strains will already be available.

On the basis that 1 ha of plantation can produce 10–12 dry tonnes of fuel each year, a 10-MW power plant will require around 7000 ha dedicated to its use. Where is this land to come from? It was initially assumed that biomass would be grown on marginal cropland or waste land. Recent experimental work has suggested that this strategy is unlikely to prove cost effective and that energy plantations require good arable land. In both Europe and North America there are now sizable areas of arable land that are set aside either because of overproduction from modern crops or for environmental reasons. This land could be used for energy crops without affecting levels of food production and while maintaining a diverse and attractive environment.

In other parts of the world agriculture is not so highly developed and levels of production are much lower. In these regions it must be considered dangerous to divert agricultural land for energy crops. Thus biomass-fired power generation based on energy plantations is most likely to mature first in the developed countries of the world.

Biomass energy conversion technology

There are a number of technologies available to convert biomass into electricity. The most widespread is a simple combustion furnace and boiler, similar to that used in a coal-fired power station. This technique, generally called *direct firing*, is a relatively inefficient method of biomass utilisation.

A second technique, biomass gasification, offers the prospect of considerably higher efficiency. This technology is currently in the development and demonstration stage and is more expensive than direct firing. An intermediate approach is to burn a small proportion of biomass mixed with coal in a coal-fired power station, a process known as *co-firing*.

There are also a number of other more specialised techniques. These include biomass digesters for converting animal wastes into a combustible gas. Biomass can also be converted into liquid fuel which can be burned in internal combustion engines.

Direct firing

The direct firing of biomass involves burning the fuel in an excess of air inside a furnace to generate heat. Aside from heat the primary products of the combustion reaction are carbon dioxide and a small quantity of ash. The heat is absorbed by a boiler placed above the main furnace chamber and water in tubes within the boiler is heated and eventually boiled, producing steam which is used to drive a steam turbine.

The simplest type of direct-firing system has a fixed grate onto which the fuel is piled and burned in air which enters the furnace chamber from beneath the grate. This type of direct-firing system, called a *pile burner*, can burn wet and dirty fuel but its overall efficiency is only around 20% at the best. The fixed grate makes it impossible to remove ash except when the furnace is shut down, so this type of plant cannot be operated continuously.

An improvement over the pile burner is the stoker combustor which has a moving grate or stoker. The moving grate allows ash to be removed continuously and fuel can be spread more evenly than in a pile burner, encouraging more efficient combustion. Air still enters the furnace from beneath the grate and this airflow cools the grate. Thus the airflow determines that maximum temperature at which the grate and hence the furnace can operate and this in turn determines the maximum moisture content of the wood that can be burnt, since the dampest wood will require the highest

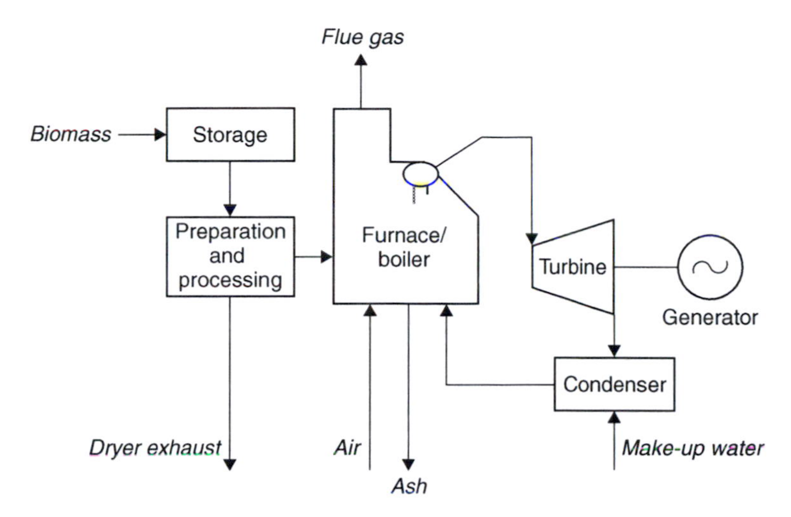

Figure 15.1 *Layout of a direct-fired biomass combustion system*

temperature if spontaneous combustion is to be maintained. There are a number of refinements to the stoker combustor such as an inclined and a water-cooled grate. Even so, maximum overall efficiency is only 25%.

Most modern coal-fired power plants burn finely ground coal which is fed into the power plant furnace through a burner and then ignites in mid-air inside the furnace chamber, a process called *suspended combustion*. It is possible to burn biomass in this way but particle size must be carefully controlled and moisture content of the fuel should be below 15%. Suspended combustion, while it can provide a higher efficiency, is not widely used in dedicated biomass power plants. However it does form the basis for co-firing which is discussed at greater length below.

As an alternative to the traditional pile burner of stoker combustor, many new biomass-fired power plants utilise a fluidised-bed furnace. The fluidised bed contains a layer of a finely sized refractory material such as sand which is agitated by passing air through it under pressure so that it becomes entrained and behaves much like a fluid. When the bed becomes hot enough, fuel mixed with the refractory bed will burn in the same way as in a conventional furnace. Fuel content within the bed in usually maintained at around 5%. Fluidised beds can burn a wide range of biomass fuels with moisture content as high as 55%. However overall efficiency is again only 25% at the best.

Direct-fired biomass power plants typically have a generating capacity of around 25–50 MW. This small size, combined with the relatively low-combustion temperature in the furnace (biomass is more reactive than coal and so tends to burn at a lower temperature) are the two main reasons for these plants' low efficiencies compared to coal plants where overall efficiencies above 40% are now common in new facilities.

Improvements are possible. Increasing the size of the typical plant to 100–200 MW will allow larger, more efficient, steam turbines to be used. New small steam turbines which incorporate advanced design features currently found only in large coal-plant turbines will also improve efficiency. Adding the ability to dry the biomass fuel prior to combustion will result in a significant increase in performance. With these changes, direct-fired biomass plants should be able to achieve 34% efficiency by the end of the first decade of this century.

Co-firing

More efficient conversion of biomass into electricity can be achieved quite simply and on a relatively large scale in another way, by the use of co-firing. Co-firing involves burning a proportion of biomass instead of coal in a coal-fired power plant. Since most coal stations operate at a much higher efficiencies than traditional direct-fired biomass plants, co-firing can take advantage of this to achieve 35–40% conversion efficiency, possibly higher in a modern high-performance coal-fired facility.

Up to 2% of biomass can be added to the coal in a coal-fired power station without any modification to the plant. The fuel is simply mixed with the coal, prepared with the coal in the plant fuel processing system and then burned in the furnace.

Above 2%, modifications are necessary. In a pulverised-coal plant these will normally include a dedicated biomass fuel processing system and changes to one or more furnace burners so that they can burn the biomass once it has been reduced to fine particles. With these changes, which are still very cheap compared to the cost of a new biomass power plant, 5–15% biomass can be burned in the furnace alongside the coal. Tests have suggested that in fact up to 40% biomass co-firing is possible but such high levels are likely to be more difficult to manage, so 15% probably represents the optimum.

Most forms of biomass, including biomass wastes and energy crops, are suitable for co-firing. Coal plants have typical unit sizes up to 600 MW, where 15% co-firing would provide 90 MW of biomass capacity. This will, in most cases, be considered 90 MW of green generating capacity. Further, biomass has a much lower sulphur content than most coals so co-firing can also reduce sulphur emissions.

Many biomass fuels, but particularly straw, have a high alkali content and this can cause problems of fouling in coal-plant boilers. Additionally, while the ash from coal-fired power plants is often used in the building industry, when biomass ash is added, the resulting residue may not have the required permit for such use. Both problems should be simply soluble.

As an alternative it is possible to combine a biomass gasifier (see below) with a coal-fired power plant. Biomass is first converted into a combustible gas and this gas is burned in the coal-plant furnace alongside the normal coal fuel. This avoids both ash and fouling problems, but at a significantly higher cost.

From an environmental point of view the primary criticism of co-firing is that the technique is so simple and cheap that it could become the principal method of achieving green energy targets where these become mandatory. This would then divert investment from other renewable technologies, damaging their development. There is no evidence yet of this happening.

Biomass gasification

For dedicated biomass combustion to become a major source of electricity, higher-efficiency conversion is required. The best means of achieving this may well prove to be biomass gasification.

Gasification involves the partial combustion of biomass in either air or oxygen to produce a gas that contains combustible organic compounds, carbon monoxide and hydrogen. The gasification process is well tested and the product gas will have a calorific value of between one-fifth and one-half that of natural gas. This is sufficient for it to be burned in a gas turbine to generate electricity.

The most efficient method of generating power using biomass gasification is to integrate the gasification plant with a combined cycle power plant. Gas from the gasifier is burned in a gas turbine and then waste heat from

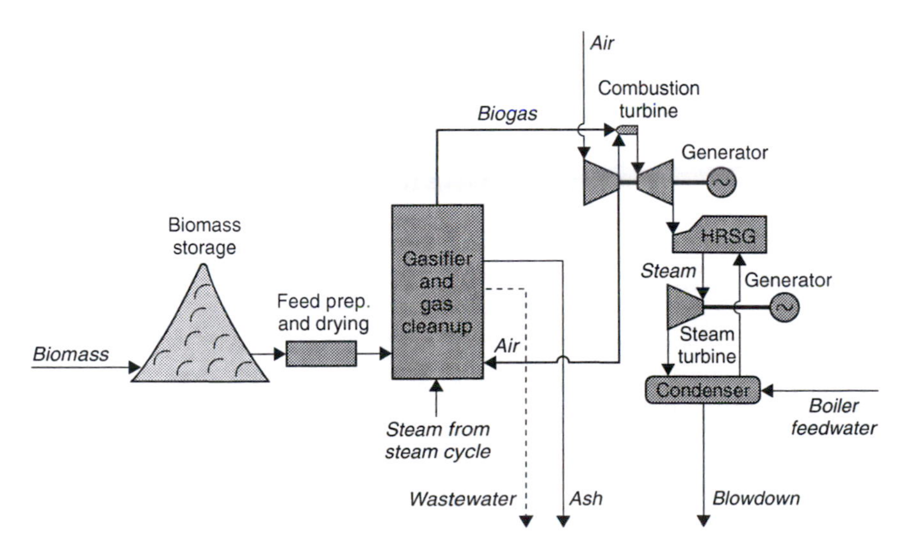

Figure 15.2 *Layout of a biomass-gasification combined cycle plant*

the turbine is used to raise steam for a steam turbine which generates additional power. Any remaining low-quality heat that cannot be used for power generation is utilised for fuel drying and fuel heating in the input stages of the plant. With this type of configuration an overall fuel-to-electricity conversion efficiency of 45% should be achievable.

An alternative approach is to use the biogas in a fuel cell. Modern, high-temperature fuel cells are capable of burning the chemical components of the biogas. While this might potentially offer even higher efficiency, perhaps 50%, the fuel cell technology is not yet commercially available.

Biomass digesters

When organic material is allowed to decay naturally but in the absence of oxygen one of the products of the process is a gas which is rich in methane. The process, called *anaerobic decomposition*, occurs when urban waste is buried in landfill sites. Significant quantities of methane are released from such sites and this gas can be collected and burned in a gas engine to generate electricity. (In many countries the collection of this gas is now mandatory since methane is a more potent greenhouse gas than carbon dioxide.)

Animal wastes from dairy and pig farms can be treated in a specially designed digester to achieve similar results, anaerobic decomposition and a methane-rich gas. The economic viability of such a scheme depends on the size of the farming operation but if the latter is large enough it can generate enough revenue to more than offset investment cost. Sewage farms which deal with human waste can also utilise similar technology.

All these applications are relatively small scale with generating units rarely larger than 100 kW, though landfill gas power plants may reach 20–30 MW. The cost effectiveness will depend in part of prevailing environmental regulations. Where these require wastes, effluents or the landfill methane gas to be collected and treated, the additional cost of a generating unit will easily be offset by the value of the power it produces. If treatment is not mandatory, other forms of disposal may prove more cost effective.

Liquid fuels

Biomass is already used extensively to produce liquid fuels. The most important of these are ethanol, made from the fermentation of grain or sugar cane and biodiesel produced from oil-rich crops such as sunflower and oil-seed rape.

Biodiesel is produced extensively in Europe where the total production was over 850,000 tonnes in 2001. Production is now on target to reach 2% of the liquid fuel market by 2005 though 5.75% by 2010 may be harder to attain.

Europe also produces ethanol but the largest producers are the USA and Brazil where the product of fermentation is added to petrol. US production is around 4 billion litres each year. Up to 10% ethanol can be blended with petrol, a product sold in the USA as gashol.

Both ethanol and biodiesel are important as substitutes for fossil-oil-derived products and their production is likely to increase both for environmental and for security reasons. Much of this fuel will be used for transportation but it is quite feasible to operate stationary engines designed to generate electricity on these fuels too. However this is likely to make only a very small contribution to global electricity generation in the foreseeable future.

Environmental considerations

Biomass can be viewed as a direct replacement for fossil fuels, particularly coal. In power generation applications it will be burnt or gasified in an entirely analogous manner to coal and like coal it will produce atmospheric emissions, principally carbon dioxide. Why, then, is biomass considered renewable?

The difference lies in the fact that biomass is a replaceable fuel. Fossil fuels such as coal and oil were originally biomass, biomass that as a result of age and geological changes had become trapped within the earth's crust. When these materials are extracted and then burned they release their carbon (their principle component) which was previously sequestered within the earth into the atmosphere. Since the industrial revolution this has led to a steady but accelerating increase in the concentration of carbon dioxide within the earth's atmosphere.

When biomass is burned it too releases carbon in the form of carbon dioxide into the atmosphere. However when replacement fuel is grown it takes carbon dioxide from the atmosphere during photosynthesis. Thus over a complete cycle of growth, harvesting and combustion there is no net addition or subtraction of carbon from the atmosphere. So burning biomass instead of coal can help stabilise carbon dioxide concentrations in the atmosphere.

Aside from carbon dioxide, combustion of biomass releases some organic compounds, carbon monoxide, particular material and nitrogen oxide into the atmosphere in exactly the same way that combustion of coal produces these materials. Depending on emission regulations these may have to be removed from a power plants's flue gas.

There is one major difference between biomass and coal. Biomass contains virtually no sulphur. Thus whereas most coal-fired power plants require an expensive flue gas scrubbing system to remove sulphur dioxide before it is released to the atmosphere, the flue gas from a biomass plant does not require this treatment.

Biomass contains virtually no toxic metals either, so the release of these into the atmosphere is reduced where biomass is burned instead of coal. Biomass also produces significantly less ash than coal, and the ash which is produced can often be returned to the soil as a fertiliser.

Life-cycle assessment

Plant emissions represent one way of assessing the environmental impact of a combustion power plant. Another way of looking at the impact is to assess the total amount of energy a plant uses to produce each unit of electricity it generates. When the results of such assessments – called *life-cycle assessments* because they look over the complete lifetime of the plant – for fossil-fuel- and biomass-fired power plants are compared, the results are unexpected.

A basic life-cycle assessment will consider all the coal or gas burnt by a fossil fuel power plant as energy consumed. However it will not consider biomass burnt in a biomass plant as energy consumed because over the lifetime of the plant all the biomass is assumed to be replaced. Such as assessment, not unexpectedly, shows a biomass plant using up to 100 times less energy to generate a kilowatt-hour of electricity[3] than a coal- or gas-fired plant.

A more realistic comparison is obtained when the fuel consumed by the fossil fuel power plant is excluded too. Typical figures for such an assessment show a direct-fired biomass power plant consuming 125 kJ to generate each kilowatt-hour, a biomass gasification power plant utilising an energy crop using 231 kJ/kWh, a coal-fired power plant using 702 kJ/kWh and a gas-fired power plant using 1718 kJ/kWh.[4] So even under these conditions the biomass power plant produces more electricity for each unit of energy it uses that the fossil-fuel-fired plant.

The reason for this is to be found in to cost of mining and transporting the fossil fuel. This is a more energy-intensive process than harvesting and transporting biofuel. The gas plant performance is also degraded by gas losses during transportation. But the results indicate that purely on an energy basis the biomass power plants are more efficient than fossil fuel plants.

Energy crops

The widespread use of energy crops for power generation will carry their own environmental implications. One of the most important of these is land use. There is a danger that the use of arable land for energy plantations will reduce that available for growing food. Currently this does not appear to be a problem in Europe and North America where sufficient land is available. In other parts of the world any significant shift in land usage could have a significant effect of food production. Under current circumstances this must be considered detrimental.

The effect of the energy plantations themselves on the environment could be beneficial, though this will depend to a large extent on how well they are managed. Most energy crops will remain in place for a number of years, improving the stability of the local environment. This can help stabilise soil conditions where erosion has become a problem and the raising of woody crops can improve groundwater retention and reduce damaging run-off of rainwater. Energy crops generally require less fertiliser than food crops, again a benefit to the environment.

Waste fuels

The use of biomass waste for power generation can also be beneficial, with the caveat already expressed that sufficient organic material is returned to the soil to maintain fertility. In California, where a significant biomass power generation capacity has evolved since the 1970s, the industry has inadvertently become a major part of the US state's waste management system. The economic value of the removal and beneficial destruction of waste in this way is not usually recognised. Were it to be taken into account, it could make the argument for biomass power generation stronger still.

Financial risks

Biomass power generation is still a relatively small sector of the power generation industry at the beginning of the twenty-first century. Much of the existing capacity is relatively inefficient and the future success of biomass as a source of electricity will depend on the development of more efficient systems for exploiting the fuel.

Improvements to traditional direct-firing technologies together with the development of biomass gasification systems are underway. New direct-firing technologies offering higher efficiencies should be available by the end of the first decade of this century. These are based on well-understood coal-plant technologies and the economic risks associated with the introduction of such techniques should be minimal.

Co-firing is a new technique which may pose a slightly higher level of risk in the early stages of its introduction. However the addition of co-firing capability to a coal-fired power plant will normally be part of a sustainable or green energy policy and the economic benefits of this – or the penalties associated with failing to implement such policies – will normally outweigh the technological risk.

Biomass gasification is in the development and demonstration stage. As a new technology it can be expected to be less reliable in its early stages of use. However it can take advantage of parallel development of coal gasification technology and this should help reduce development costs and help improve reliability. Commercial biomass gasification plants should be available during the second decade of the century.

Agricultural risk

As outlined above, the technological risk associated with biomass power generation is relatively low and predictable. However biomass power generation requires both power plants and fuel; the fuel end of the equation poses a much higher level of risk.

A successful biomass power project will have to become closely integrated with the agricultural production of the fuel. However there is, today, no agricultural industry devoted to the raising of energy crops. Crop raising expertise, harvesting expertise and a transport infrastructure must all be established. Yet these cannot be established until there is a power industry to buy the crop they produce. Here lies the difficulty.

If this vicious circle is to be broken, some form of subsidy will almost certainly be necessary. Small programmes already exist in countries such as the UK but other and larger programmes will be needed. Subsidies are themselves unpredictable, capable of being withdrawn or phased out at short notice. Thus the risk of fuel supply failure must be considered high.

Over the longer term the energy crop industry will be in a position to offer long-term fuel supply contracts to power plant owners. Until that happens, each project will have to work closely with a fuel supplier to ensure it has a viable future.

Biomass waste fuels remain as a standby but supplies of these can be unpredictable and if demand outstrips supply, fuel costs can escalate. Such a situation occurred in California during the 1990s, forcing a number of biomass power plants to shut down operations. So until an energy crop industry has been established, fuel will remain the weak link in the biomass power supply chain.

The cost of biomass generated power

A biomass-fired power station is technically similar to a coal-fired power plant and the economics of the two are based on similar principles. In both cases the cost of the electricity generated depends on two factors, the cost of the plant and the cost of the fuel.

Technology costs

The cheapest option for generating electricity from biomass is co-firing. Retrofitting a co-firing option to an existing coal-fired power plant costs between \$100/kW and \$700/kW of biomass generating capacity, with the average price around \$200/kW.[5]

A new direct-fired biomass power plant costs around \$2000/kW as a result of the low efficiency of existing technology. This could be reduced to around \$1300/kW by the end of the decade as new, higher-efficiency plants are introduced. First generation biomass gasification power plants will probably cost around \$2000/kW, dropping to \$1400/kW by 2010. For comparison, a typical coal-gasification-based plant entering service in 2006 would cost around \$1300/kW.

Fuel costs

Biomass waste fuel costs vary widely but estimates across the USA suggest that the cost in North America is between \$25/dry tonne and \$60/dry tonne. Sawmill waste used close to the process plant might be obtained even more cheaply, for perhaps \$17/dry tonne.

More important from a long-term perspective is the cost of energy crops. Table 15.3 shows some typical costs from around the world, but these figures should be used as a broad guide only, since the methods by which they were evaluated are not all directly comparable. As the table shows, costs can vary between \$1/GJ and \$5/GJ. Coal costs around \$2/GJ.

In the USA, experimental projects suggest that switchgrass might be delivered for as little as \$2.1/GJ and wood for \$3.3/GJ. On this basis, switchgrass would be almost competitive with coal, but wood does not appear to be competitive yet.

Electricity costs

Experience from California during the 1990s indicates that biomass plants were able to operate profitably when the wholesale cost of electricity was

Table 15.3 *Biomass fuel costs*

	Cost (\$/GJ)
Brazil (northwest)	0.97–4.60
China (southwest)	0.60
Hawaii	2.06–3.20
Portugal	2.30
Sweden	4.00
USA	1.90–2.80

Source: US Department of Energy (Oak Ridge National Laboratory).

$0.040/kWh; they did, however benefit from a subsidy of $0.015/kWh, suggesting that they could generate power for $0.055/kWh. More generally, a plant with an efficiency of around 23% could deliver power at $0.05/kWh provided the cost of the fuel was below $1/GJ, or well below the cost at which fuel will be available in the USA in the foreseeable future. However if the plant has an efficiency of 35%, the fuel cost could rise to $2.8/GJ. It should be possible to deliver an energy crop for this price in the USA today but the energy conversion efficiency required is only currently available with co-firing.

These estimates apply to the USA and even there they can only offer broad guidance. However it seems probable that co-firing could deliver power at a competitive price today in some parts of the world whereas dedicated biomass power plants will not be able to compete effectively without some form of incentive.

End notes

1 Energy Technology – The Next Steps, published by the EU Directorate General for Energy in December 1997.
2 World Energy Council, Survey of Energy Resources, Biomass, 2001.
3 Biopower Technical Assessment, State of the Industry and the Technology, Richard L. Bain, Wade P. Amos, Mark Downing and Robert L. Perlack (January 2003) (NREL/TP-510-33132).
4 Refer *supra* note 3.
5 Refer *supra* note 3.

16 Power from waste

The generation of power from waste is a very specialised industry and its principal aim is not to produce electricity. Power-from-waste plants are combustion plants designed to destroy or reduce in volume municipal and in some cases industrial waste.[1] As an incidental, but nevertheless valuable by-product, the processes adopted to manage these wastes may also be capable of generating electricity.

The level of exploitation of waste-to-energy plants varies from country to country. They have been used widely in parts of Europe, where waste has been burned since the end of the nineteenth century, and form a major part of Japan's waste disposal strategy. In contrast the USA has only adopted the technology patchily. In addition environmental concerns about the emissions from the plants has caused recent resistance to their construction both in the USA and elsewhere.

Where they are employed, these plants generally burn domestic and urban refuse – called in this context *municipal solid waste* (MSW) – using the resulting heat to generate steam to drive a conventional steam turbine. MSW can also be sorted and treated to produce a compacted fuel called *refuse-derived fuel* (RDF) which can be burned in a power station.

Some industrial waste may be treated in the same way. However industrial wastes are likely to contain toxic materials which have to be handled using special procedures. Where such care is not required, they can be dealt with in the same way as urban waste.

There are a number of other categories of waste, primarily resulting from the agricultural and forestry industries, that can be used to generate electricity. These have been dealt with under biomass in Chapter 15, which also dealt with the collection and use of methane produced in landfill refuse disposal sites. However we need to consider landfill briefly here since it offers the main alternative to waste combustion.

Landfill waste disposal

The landfill site – essentially and enormous hole in the ground where waste is dumped – is the main alternative to the technologies discussed in this chapter as a means of waste disposal. Though crude, its simplicity has led to it becoming the favoured method of urban waste disposal across the globe.

While landfill use remains popular in many countries, it is coming under pressure in others. This is partly a result of the demand for land which increasingly restricts that available for waste burial. More potent still are environmental concerns about the long-term effects of landfill disposal, effects resulting from the methane emissions from such sites (discussed in Chapter 15) and from the seepage of toxic residues into water supplies.

Such concerns have already led the European Union (EU) to legislate[2] to restrict the use of landfill waste disposal. Similar legislation is bound to follow in other parts of the world. But waste will still be produced. This is where technological solutions, such as the power-from-waste plant, enter the equation.

Power-from-waste technology is not cheap. The specialised handling that waste requires, coupled with the need for extensive emission-control systems to prevent atmospheric pollution, make such plants much more expensive to build than any other type of combustion power plant. They are also expensive to operate.

If these plants had to survive on the revenue from power generation alone, they would never be built. Fortunately they have another source of income. Since waste has to be disposed of in a regulated manner, waste disposal plant operators can charge a fee – normally called the *tipping fee* – to take the waste. The tipping fee represents the main source of income for a power-from-waste plant. Any additional income derived from power generation will benefit the economics but the plant may well be able to survive without it.

Waste sources

There are two principle types of waste suitable for disposal in a power-from-waste plant: urban (primarily domestic) refuse, normally referred to a MSW, and industrial waste. Some industrial waste is broadly similar in content to MSW and this can be treated in the same way as the latter. Other industrial waste must be dealt with differently because of the hazardous or valuable materials it contains. This chapter is only concerned with MSW and it will not deal with industrial waste except where it can be burned with MSW.

The main source of MSW is an urban community.[3] The quantity and size of such communities is growing rapidly. In the last two generations the number of people living in cities has increased by between 250% and 500%.[4] This has been particularly notable in the developing world where the number of urban dwellers is expected to reach 2.71 billion by 2010. A further 1 billion live in the cities of the developed world. Thus, close to half the population of the world will be living in cities by 2010.

Urban dwelling has grown, particularly, rapidly in South America and the Caribbean where, by 2025, 80% of the populations will be living in towns.

But these regions are not unique. Urban communities are growing virtually everywhere. These towns and cities constitute the source of MSW.

The amount of waste these populations produce varies from country to country and from continent to continent. In general, the city dweller in an industrialised country produces far more waste than one in a developing country. Thus a typical Californian might produce 1.3–1.4 kg each day while a city dweller in Mexico City produces only half that. A Nigerian town dweller probably produces less than 200 g of waste each day.

In the mid-1990s the International Energy Agency estimated that developed countries alone produced an estimated 426 million tonnes of waste each year. If all this was used to generate electricity, potential output would be 191 TWh/year. Annual energy demand in 2001 was 13,290 TWh.[5]

Waste composition

The composition of the waste varies from place to place. In general the waste from the urban household in an industrialised country will contain 30–40% paper and cardboard and up to 10% plastic. The proportions of these in the waste from a household in the Dominican Republic will be much lower but the Dominican household's waste will probably contain 80% food waste whereas the proportion in a US household waste may only be 26%.[6]

There are other important differences. The waste from households in developing countries contains a high proportion of moisture, often as high as 50%, making it difficult to burn without first reducing the moisture content by drying. In contrast, the high proportions of paper and plastic in the waste from a household in the industrial world make it much easier to burn.

All these factors affect the energy content of waste, and energy content is a crucial factor in determining the viability of a power-from-waste plant. Unless the plant can produce enough excess heat from waste combustion to raise steam, it cannot expect to generate any electricity.

Table 16.1 provides some figures for MSW energy content from different parts of the world. US waste has the highest-energy content, 10,500 kJ/kg, approaching that of sub-bituminous coal (see Table 16.1). European cities

Table 16.1 *Energy content of urban wastes from different regions*

	Energy content (kJ/kg)
USA	10,500
Western Europe	7500
Taiwan (Taipei)	7500
Mid-sized Indian cities	3300–4600
Sub-bituminous coal	10,700–14,900

Source: United States Agency for International Development.[7]

and prosperous Asian cities such as Taipei generate waste with around 7500 kJ/kW. The waste from typical mid-sized Indian cities contains roughly half this amount of energy.

In the latter case the low-energy content may not be entirely due to the quality of waste. In cities in India – but not them alone – much of the urban waste is collected by city sweepers. Such waste is contaminated with considerable quantities of stone, earth and sand. In Bombay, for example, the amount of non-combustible material of this type in waste may reach 30%. Not only does this reduce the energy content of the waste, it could also damage a combustion system so the design of a waste disposal plant has to take its presence into account.

Given such local variations in waste content it is vitally important, before a power-from-waste plant is built, that the waste available be carefully assessed. For that, local waste-collection procedures and organisations have to be examined.

Waste collection

Urban refuse collection is organised in different ways in different parts of the world. In some countries it is run by municipalities, in others it is provided by private operators. Where a municipality run waste collection as a service, the same city might build and operate its own power-from-waste plant. Under these circumstances the composition of the waste can be readily assessed and controlled if necessary.

More often waste collection is carried out by private companies. The waste that these companies provide will vary in quality. In some cases it will contain the whole range of waste, but in others it will have been sorted to remove the more valuable material. Some countries now require that glass, metal, plastic and paper be recycled. This too will affect the quality of the MSW available.

Inevitably the quality of waste will vary by season. Economic factors are also important. Waste will be poorer in a recession than in a boom. Local variations can also be significant. Richer neighbourhoods tend to produce better quality waste than poorer neighbourhoods. This has led to the suggestion that the quality of waste for a power-from-waste plant might be maintained by collecting only from prosperous areas of a city.

Whatever the strategy, knowledge of the waste, its source and its variations will form a necessary part of the management of a waste-to-energy plant. That information can only be gained with practical experience, by analysis of waste collected by the contractor that will provide waste for the plant. Even with this knowledge, it may be impossible to maintain an adequate energy content in the waste throughout the year. Then the only solution may be to add some higher-energy content fuel to the waste. Biomass waste from local sources will often be the most economical solution in this situation.

Waste power generation technologies

A power-from-waste plant is a power station fuelled with urban waste. As already indicated, such a facility may have as its primary function, waste disposal. Nevertheless the technologies employed will be traditional power generation technologies as used in combustion plants. Combustion systems include grate burners, some fluidised-bed burners, and more recently gasification and pyrolysis. Heat generated in these combustion systems is used to raise steam and drive a steam generator.

Within the broad outline above, power-from-waste plants vary enormously. Much depends on the waste to be burnt, its energy content, the amount of recyclable material or metal it contains and its moisture content. Waste may be sorted before combustion or it may be burnt as received. Emission-control systems will vary too, with toxic metals and dioxins a particular target, but nitrogen oxides, sulphur dioxide and carbon monoxide emissions must all fall below local limits. Carbon dioxide emissions may need monitoring to comply with greenhouse gas emission regulations.

Once the waste has been burnt, residues remain. Power-from-waste plants will generally reduce the volume of waste to around 10% of its original. A way must then be found to dispose of this residual ash. If it is sufficiently benign, it may be used as aggregate for road construction. Otherwise it will probably be buried in a landfill. Other residues from emission-control systems will have to be buried in controlled landfill sites too.

Northern Europe has been the traditional home of waste incineration plants for power generation. Altogether there were around 250 municipal waste combustion plants in the EU in the late 1990s, most in the northern countries of the Union. Between them they had a generating capacity of around 1500 MW, almost half the global total of 3200 MW in 1997.[8] Japan has also made extensive use of waste combustion, though not always for power generation. In 1999 there were about 600 waste-to-energy plants in operation worldwide.

Europe has also developed the most widely used waste combustion technology. Two companies, Martin GmbH based in Munich and the Zurich company Von Roll, accounted for close to 70% of the market for the dominant technology, called *mass burn*, at the turn of the century.[9] The rest of the market is divided among a number of smaller companies, most based in either Europe, the USA or Japan.

The dominant European technology has been widely licensed. It was the source of the technology used in most US power-from-waste plants built in the late 1970s and early 1980s. More recently several developing countries of Asia have taken interest in power from waste and European technology has been modified for use in China.

Newer technologies based on gasification and pyrolysis are being developed by a variety of companies. These are based on technologies from other industries such as power generation and petrochemicals.

Traditional combustion plants

The traditional method of converting waste to energy is by burning it directly in a special combustion chamber and grate, a process which is often called mass burning. The dominant European technologies use this system. These involve specially developed moving grates, often inclined to control the transfer of the waste, and long combustion times to ensure that the waste is completely destroyed. Designs have evolved over 20–30 years and are generally conservative.

More recently, fluidised-bed combustion systems have sometimes been used in place of traditional grates. Such systems are good at burning heterogeneous fuel but they require the waste to be reduced to small particles first. These systems remain relatively rare.

The actual grate forms only a part of a waste treatment plant. A typical solid-waste combustion facility is integrated into a waste-collection infrastructure. Waste is delivered by the collecting trucks to a handling (and possibly a sorting) facility where it must be stored in a controlled environment to prevent environmental pollution. Recyclable materials may be removed at this stage, though metallic material is often recovered after combustion. Grabs and conveyors will then be used to transfer the waste from the store to the combustor.

Plant components, and particularly the grates, must be made of special corrosion-resistant materials. The grate must also include a sophisticated combustion-control system to ensure steady and reliable combustion while the quality and energy content of the refuse fuel varies. In some more modern systems oxygen is fed into the grate to help control combustion. The temperature at which the combustion takes place must usually be above 1000°C to destroy chemicals such as dioxins but must not exceed 1300°C as this can affect the way ash is formed and its content.

Hot combustion gases from the grate flow vertically into a boiler where the heat is captured to generate steam. The combustion process in the grate and the temperature profiles within the boiler have to be maintained carefully in order to control the destruction of toxic chemicals. Most of the residual material after combustion is removed from the bottom of the combustion chamber as slag. However there may be further solid particles in the flue gases, some of which can be recycled into the furnace.

Upon exiting the combustion and boiler system, the exhaust gases have to be treated extensively. While the combustion chamber may utilise techniques to minimise nitrogen oxide emissions – though further reduction may prove necessary – a system to capture sulphur will be required. This will probably be designed to capture other acidic gases such as hydrogen chloride too. There may be a further capture system based on active carbon which will absorb a variety of metallic and organic residues in the flue gases. Then some sort of particle filter will be needed to remove solids. By this stage the exhaust gases should be sufficiently clean to release into the

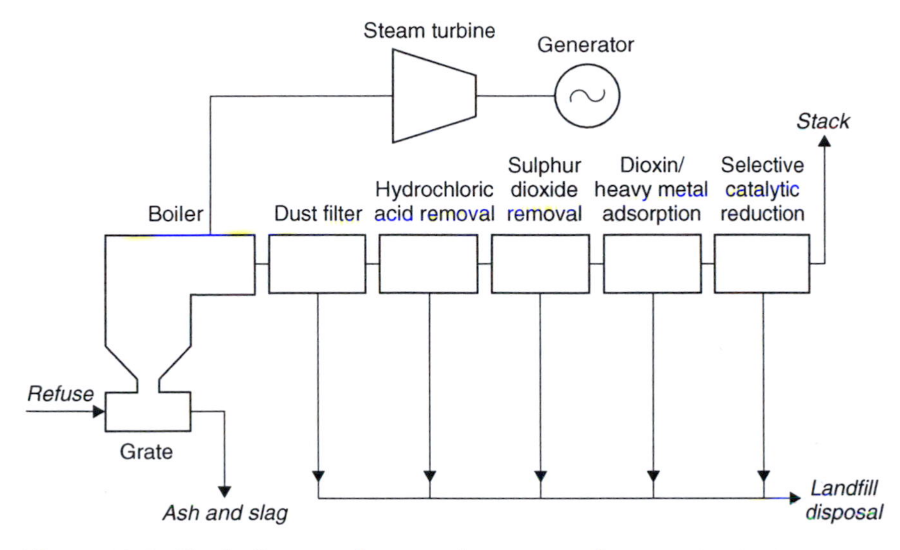

Figure 16.1 *Block diagram of a mass-burn power-from-waste plant*

atmosphere but continuous monitoring systems are required to make sure that emission standards are maintained.

Dust from the flue gas filters is normally toxic and must be disposed of in a landfill. Other flue gas treatment residues will probably need to be buried too. The slag from the combustor may, however, be clean enough to exploit for road construction. Modern mass-burn plants aim to generate slag that can be utilised in this way.

Mass-burn plants may burn up to 2000 tonnes/day of MSW. Where a smaller capacity is required, a different type of combustion system, called a *rotary kiln*, can be employed. As its name suggests, this system uses a rotating combustion chamber which ensures that all the waste is burned. The chamber is inclined so that the material rolls from one end to the other as it burns. Such combustors are capable of burning waste with a high-moisture content, perhaps up to 65%. Capacities of rotary kilns are up to 200 tonnes/day of refuse, suitable to meet the needs of small urban communities.

Gasification and pyrolysis

In recent years a number of companies have attempted to develop new waste-to-energy technologies based on both gasification and pyrolysis. These technologies are derived from the power and the petrochemicals industries.

Pyrolysis is a partial combustion process carried out at moderate temperatures in the absence of air, which usually produces a combustible gas and a combustible solid residue. Gasification uses higher temperatures and converts most of the solid material into a combustible gas. In both cases the gas will normally be burnt to generate heat and thence steam.

Typical of this type of plant is a system developed in the 1990s in Japan[10] which employs an initial pyrolysis process followed by combustion to generate heat. Waste delivered to the plant is first shredded and then fed into a rotating pyrolysis drum where it is heated to around 450°C. The heat, provided by hot air generated at a later stage in the process, pyrolyses the waste, converting it into a combustible gas and a solid residue.

The solid residue contains any metal which entered with the waste. This can be removed at this stage for recycling. Both iron and aluminium can be segregated in this way. The remaining solid slag is crushed. The gas and the crushed residue are then fed into a high-temperature combustion chamber operating at 1300°C where it is completely burnt. Combustion is controlled to limit nitrogen oxides formation. Incombustible material adheres to the walls of the combustion chamber where it flows, in liquid form, to the bottom. From here it is led out of the bottom of the furnace and immediately quenched, creating an inert granular material suitable for road building.

Hot flue gases from the combustion chamber are used to generate steam to drive a turbine. Dust is then removed from the exhaust gases and returned to the combustion system. Following this, a flue gas treatment system removes any remaining acid gases. Only this material, around 1%

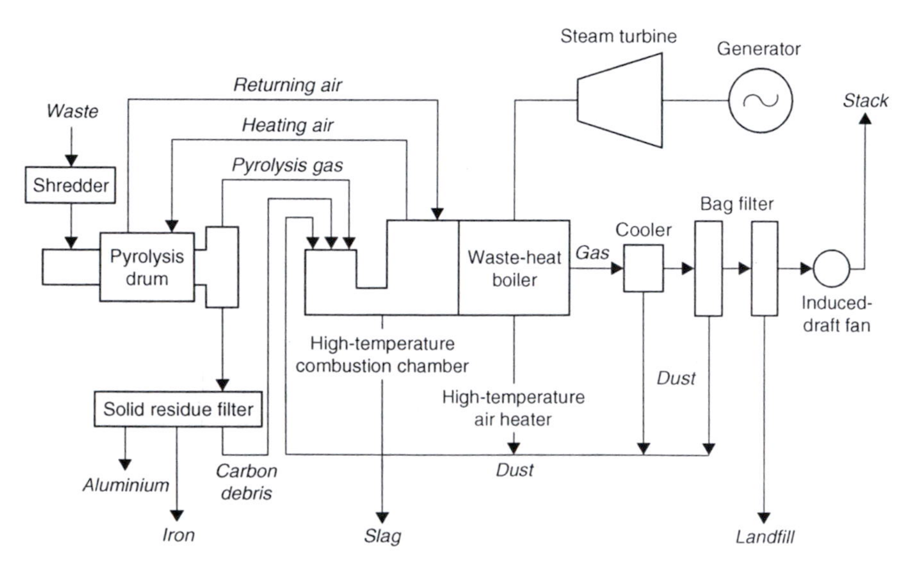

Figure 16.2 *Block diagram of a waste pyrolysis plant*

of the original volume of the MSW, needs to be disposed of in a landfill. The system also claims to keep residual levels of dioxins extremely low.

Waste gasification is similar to pyrolysis but conversion of waste takes place at a higher temperature in the presence of a controlled amount of air or oxygen. Depending on the process used a low- or medium-energy content synthetic gas will be produced. In a power-to-waste plant this will be burnt but it can also be used as a feed for some chemical processes or as a means of generating hydrogen.

Refuse-derived fuel

RDF is the product of the treatment of MSW to create a fuel that can be burnt easily in a combustion boiler. In order to produce RDF, waste must first be shredded and then carefully sorted to remove all non-combustible material such as glass, metal and stone. Shredding and separating is carried out using a series of mechanical processes which are energy intensive. The World Bank has estimated that it requires 80–100 kWh to process 1 tonne of MSW and a further 110–130 kWh to dry the waste.[11]

After the waste has been shredded and separated, the combustible portion is formed into pellets which can be sold as fuel. The original intention of this process was to generate a fuel suitable for mixing with coal in coal-fired power plants. This, however, led to system problems and the modern strategy is to burn the fuel in specially designed power plants. An alternative is to mix the RDF with biomass waste and then burn the mixture in a power plant. Since RDF production must be preceded by careful sorting, this type of procedure is best suited to situations where extensive recycling is planned.

Environmental considerations

Urban waste, its production and its fate are major environmental issues. Modern urban living produces enormous quantities of waste in the form of paper, plastic, metals and glass as well as organic materials. How these wastes are processed is a matter of increasing global concern.

Wastes such as paper, glass and metal can be recycled, as can plastics in theory. From an environmental perspective it makes sense to reuse as much waste as possible, so environmentalists generally favour maximum recycling. Many European governments promote recycling. However the economics of recycling are not clear cut and there are critics who consider it ineffective. Since such debates pitch sustainability against economy, the issue is not easily resolved.

While recycling offers the ideal solution, in practice there are often neither the facilities nor the infrastructure to recycle effectively. Even where recycling is employed there is still a residue of waste that cannot be reused. Thus there remains a considerable volume of waste for which an alternative means of disposal is required. The only options currently available are burial in a landfill site or combustion.

The combustion of waste would seem initially as the ideal solution. Combustion reduces the quantity of waste to 10% or less of its original volume. At the same time it produces energy as a by-product and this energy can be used to generate electricity or for heating, or both. Unfortunately waste often contains traces of undesirable substances which may emerge into the atmosphere as a result of combustion. Other hazardous products may result from the combustion itself, with the waste providing the chemical precursors. So, while solving one environmental problem, waste combustion can generate others.

In the face of this, the combustion of waste is becoming increasingly subject to strict legislation. This sets limits on amounts of different hazardous materials which can be released as a result of the process. Chief among these are heavy metals such as mercury and potent organic compounds such as dioxins. Modern waste-to-energy plants appear able to meet these requirements. However they have acquired a bad reputation in the past 20 years in some parts of the world. This has proved difficult to overcome and there are countries where power-from-waste plants are considered too unpopular to gain approval. New waste conversion technologies such as gasification and pyrolysis may be able to breach this barrier.

Waste plant emissions

A plant burning waste produces four major types of product. Firstly there is a solid residue from the grate itself, normally termed *slag* or *ash*. Secondly there is a chemical product resulting from flue gas treatment systems. Thirdly there is a quantity of dust in the flue gases emerging from the plant boiler; this is normally captured with filters or an electrostatic precipitator. Finally there is the flue gas itself.

Ash

The nature of the ash or slag emerging from the grate of a power-from-waste plant will depend on both the type of waste being burnt and the combustion conditions. While its primary constituents will be solid, incombustible mineral material from the wastes, this residue will be contaminated with traces of a variety of metals. These traces may be in a toxic or a harmless form.

By careful control of the temperature in the furnace, it is possible to incorporate the metals into the mineral content of the ash and render them effectively harmless. This is a process called *sintering*. The effectiveness of the sintering process in rendering toxic metals harmless will be determined by measuring the amounts capable of being leached out by water. The ash may also contain some toxic organic compounds such as dioxins. Furnace conditions can minimise these too since a sufficiently high temperature will normally destroy such compounds. The effectiveness of this will again be determined by a leaching test.

If the ash or slag is too toxic it will have to be buried in a landfill. Modern facilities aim to render it sufficiently stable and benign that it can be used for road building or for similar purposes. When they succeed, only a residual 1% of the original waste needs to be buried.

Fly ash and flue gas treatment residues

Fine solid particles called *fly ash* escape with the flue gas from a furnace. This fly ash will often contain high levels of toxic metals and must be captured. Capture is achieved either by using a fabric filter called a *bag filter*, or by employing a device called an *electrostatic precipitator*. Both should be capable of removing close to 100% of the dust from the flue gas. Once captured this dust must be safely buried in a landfill.

The same applies to the chemical residues which result from the various flue gas treatment systems used to remove harmful material from the exhaust gases of the plant. Depending on the treatment process, the residue may be a solid or a wet slurry. In the latter case, the slurry will normally be dried using the hot exhaust gases before disposal.

Flue gas

Once treated, the flue gas from a waste combustion plant should be sufficiently clean to release into the atmosphere. The gas will usually need to be monitored continuously to ensure that emission limits are being met.

Dioxins

One of the most potent environmental concerns during the last 20–30 years has related to the release of dioxins into the atmosphere. Dioxins are undesirable by-products of the manufacture of a variety of chemicals such as pesticides and disinfectants, but one particular compound called *2,3,7,8-tetrachlorodibenzo-p-dioxin* has come to be identified as *dioxin*. This material was thought to be extremely toxic to humans, though more recent studies suggest earlier results were exaggerated.

Dioxins can be found in urban waste and there is also a danger that the compounds can be formed during waste combustion if the process is not carefully controlled. Some early waste incineration plants did not control the emissions sufficiently carefully and this led to instances of widespread contamination. Such instances have coloured the perception of waste-to-energy plants ever since.

Dioxin emission levels are now closely regulated and emissions have fallen. In the USA, the emissions of dioxins from large waste-to-energy facilities fell from 4260 g (toxic equivalent) in 1990 to 12 g (toxic equivalent) in 2000.[12] The European emission limit for dioxins is $0.1\,ng/Nm^3$. Power-from-waste plants built in the middle of the first decade of the twenty-first century should be capable of reducing the emission level to one-tenth of this.

Heavy metals

Heavy metals, particularly mercury have proved another source of concern. Less mercury is used today than in the past. This combined with better filtration systems has reduced mercury emissions from power-from-waste plants in the USA to around 2 tonnes/year. Coal-fired power plants release over 40 tonnes/year.

There are other metals such as cadmium and lead which must be monitored. However in general the emissions of metals from waste incineration plants should fall well below legal emission limits. Today proponents of these plants would argue that they are significantly less polluting than landfills. New technologies may well be able to provide even higher-emission performance. Whether this will be sufficient to overcome the reputation which has already attached itself to such plants remains to be seen.

Financial risks

The traditional technology used for waste combustion is robust and extensively tested. Any risk associated with its use is small and well documented. New technologies under development such as gasification and pyrolysis have not yet been proved and the risks associated with their use are higher.

There is also a risk associated with the waste which is to provide the fuel for a power-from-waste plant. It is important to ascertain exactly what type of waste will be available to a particular project and its typical content. This can only be discovered by careful analysis of actual samples. Long-term analysis is necessary since waste content varies seasonally. However waste quality can also vary over a longer time scale, particularly if the supplier changes or where there are demographic changes within the catchment area.

Regulations and legislation also pose a threat. Any planned project will be required to meet current regulations but these may change once the plant has been built, necessitating modifications to meet new requirements. While the legislative situation is stable in areas such as Europe, it may not be everywhere. It would seem prudent when planning a project to choose the best technology available since this is likely to meet both current and future regulations anywhere in the world.

Perhaps the greatest risk with a power-from-waste project relates to its economics. If a plant is to be operated as a public service, then the economic viability will normally be guaranteed by the public sector. If it is to be a wholly private sector project, then the viability will depend on the value to waste collectors of the service offered. The price collectors are prepared to pay will depend on the competition. Under these circumstances, long-term contracts may offer the best security.

The cost of energy from waste

The capital cost of equipment to generate electricity from waste is generally much higher than for conventional power generation equipment to burn fossil fuel. Plant design is specialised and must include refinements for emission control that are not necessary in the fossil fuel plant. Grate design is unique too.

Against this must be offset the revenue of the plant, not only from the electricity generated but also from the fuel itself, the waste. Industry and municipalities expect to pay to dispose of their waste. Consequently, the economics of a project should be designed so that the revenue from the waste disposal contracts is adequate to enable the power from the plant to be sold competitively.

The cost of a typical municipal waste combustion plant is $5000–10,000/ kW, at least three times the cost of a coal-fired power plant of the same generating capacity. Smaller plants will be relatively more expensive. The cost of operating a plant is probably three times that of a coal-fired power plant too. According to US government estimates, such plants generate electricity at between $0.02 and $0.14/kWh.

End notes

1 Modern plants often recycle any reusable material, burning only the remainder.

2 European Union, Council Directive 1999/31/EC of 26 April 1999 on the landfill of waste, Official Journal of the European Communities, pp. L182/1–19 (July 1999).

3 In the developed world the waste from rural communities is often handled in a similar way.
4 Mining the Urban Waste Stream for Energy: Options, Technological Limitations, and Lessons from the Field, United States Agency for International Development, 1996 (Biomass Energy Systems and Technology Project DHR-5737-A-00-9058-00).
5 US Energy Information Administration, International Energy Outlook, 2004.
6 Refer *supra* note 4.
7 Refer *supra* note 4.
8 This figure is from the EU DG of Energy.
9 An overview of the global waste to energy industry, Nickolas J. Themelis, Waste Management World (July–August 2003).
10 The process, called R21, was developed by Mitsui Engineering and Shipbuilding. The first plant was completed in 2000.
11 Refer *supra* note 4.
12 US Environmental Protection Agency. These figures are quoted in 'An overview of the global waste-to-energy industry' Waste Management World (July–August 2003).

17 Nuclear power

Nuclear power is the most controversial of all the forms of power generation. To evaluate its significance involves weighing political, strategic, environmental, economic and emotional factors which attract partisan views far more strident that any other method of electricity generation.

From its origins in the atomic weapons programme of the World War II nuclear power generation grew, by the beginning of the 1970s, into the great hope for unlimited global power. In 1974, the US power industry alone had ordered 200 nuclear reactors and in 1974 the US Energy Research and Development Administration estimated that US nuclear generating capacity could reach 1200 GW by 2000. (Total US generating capacity in 2002 from all sources was 980 GW.[1]) The UK, France, Germany and Japan all began to build up substantial nuclear generating capacities too.

But even as orders were being placed, the nuclear industry was reaching a watershed. A combination of economic, regulatory and environmental factors conspired to bring the development of nuclear power to a halt in the USA. Similar effects spread to other countries.

There were already environmental and safety concerns during the 1970s but two accidents, one at Three Mile Island in the USA in 1979 and a second at Chernobyl in the Ukraine in 1986, turned public opinion strongly against nuclear power. In response new safety regulations were introduced, lengthening construction times and increasing costs. By the late 1980s, 100 nuclear projects in the USA had been cancelled. To make matters worse, nuclear waste disposal had became a political issue that could not be resolved. As of 2004, no new nuclear reactor has been ordered in the USA since 1978.

The US still retains a large fleet of nuclear power stations. Some countries in Europe and Scandinavia decided to rule the option out completely. In 1978 Austria voted to ban nuclear power. Sweden voted in 1980 to phase out nuclear power by 2010, although this timetable may yet be abandoned. Germany reached an agreement with its nuclear power producers in 2000 to phase out its nuclear stations.

Other Western countries such as France, Belgium and Finland remain positive about nuclear generation. The UK government, too, retains a nuclear option. And in 2003 the Finnish utility Teollisuuden Voima Oy (TVO) ordered a new nuclear unit, the first that will have been under construction in the European Union (EU) for over a decade.

There is also a large fleet of nuclear power plants in Eastern Europe. These plants are all based on Russian-designed reactors. The safety of the

Russian designs has been a matter of concern since the Chernobyl accident in 1986. From the beginning of the 1990s, when cold war barriers fell, efforts have been made to improve the safety of Eastern European reactors or to force their closure. No new plants have been started since then.[2]

The evolution of nuclear generation in Asia has followed a different course. Japan has continued to develop its installed nuclear base, as has South Korea, though the Japanese nuclear industry began to face considerable public criticism at the end of the twentieth century. Taiwan ordered two new nuclear reactors in 1996; public pressure may make these the last that country builds. India has an indigenous nuclear industry. And in the mid-1990s, China started to develop what promises to be a strong nuclear base. These nations, but primarily China, are keeping the nuclear construction industry afloat.

Global nuclear capacity

At the end of 1999, according to figures compiled by the World Energy Council[3] there were 430 operating nuclear reactors, worldwide. (There were 437 operating in 1995.[4]) These had a total generating capacity of 349 GW. A further 41 units were under construction; these had an aggregate capacity of 33 GW.

The global figures are broken down in Table 17.1 to show the distribution of current nuclear generating capacity by region. Europe, with 215 units and 171 GW, has the greatest capacity. North America has 120 operating units with an aggregate generating capacity of 109 GW while Asia has 90 units. Of the continents, only Australia and Antarctica have none.

Nationally, France produces around 75% of its electricity from nuclear power stations. Lithuania generates 73% from nuclear sources and Belgium 58%. In Asia, South Korea produces 43% of its power from nuclear units

Table 17.1 *Global nuclear generating capacity*

	Number of units	*Total capacity (MW)*
Africa	2	1800
North America	120	108,919
South America	3	1552
Asia	90	65,884
Europe	215	170,854
Middle East	1	1000
Total	431	350,009

Source: World Energy Council.

while Japan relies on nuclear power for 12% of its electricity. In all 18 countries rely on nuclear plants for 25% or more of their electricity.[5]

Globally, these nuclear plants provide around 16% of total electricity generation, almost as much as hydropower. Net generation at the end of 1999 was 2391 TWh.[6] However the overall global nuclear capacity is now static; new plants built in Asia compensate for old plants removed from service in other parts of the world.

Economically nuclear power plants are perceived to be expensive to build, particularly in the USA. However plants where the capital cost has been written off have proved extremely competitive generators of electricity. In the USA, for example, the cost of base-load nuclear power averaged $0.0171/kWh in 2002, undercutting all other sources of electricity.[7]

The future

For the reasons already outlined, the nuclear power industry looked moribund at the end of the twentieth century in all but a handful of Asian countries. The twenty-first century has brought new hope. Against all expectations, nuclear power plants in the USA are often faring well in the deregulated electricity market and their value is increasing. This may encourage a more positive attitude towards nuclear plants within the financial sector there.

The development of new reactors that are cheaper and quicker to build and which are safer may help improve perceptions. Meanwhile global warming offers the nuclear industry an opportunity to sell its product as a zero greenhouse emission technology. This argument has not won support within the environmental lobby which still perceives nuclear power as a pariah. The industry has, however, been successful in lobbying for support within the US government which wants to build a new generation of nuclear plants. The UK government appears to hold the option of new nuclear capacity open too.

Major issues still remain if nuclear power is to be rehabilitated. The disposal of nuclear waste is a significant problem and one that appears no nearer a satisfactory solution than it did in the 1980s or 1990s. Nuclear proliferation renders nuclear power suspect because it is a source of fissile weapons material. The dangers of terrorism have also raised the safety stakes as far as the nuclear industry is concerned. These are serious issues. If concerns relating to them can be met, the nuclear industry may see the renaissance it desperately seeks. But that renaissance is far from certain.

Fundamentals of nuclear power

A nuclear power station generates electricity by utilising energy released when the nuclei of a large atom such as uranium split into smaller components, a process called *nuclear fission*. The amount of energy released by this fission process is enormous. One kilogram of naturally occurring uranium could, in theory, release around 140 GWh of energy. (140 GWh represents the output of a 1000 MW coal-fired plant operating a full power for nearly 6 days.)

There is another source of nuclear energy, nuclear fusion, which involves the reverse of a fission reaction. In this case small atoms are encouraged to fuse at extraordinarily high temperatures to form larger atoms. Like nuclear fission, fusion releases massive amounts of energy. However it will only take place under extreme conditions. Fusion of hydrogen atoms is the main source of energy within the Sun.

The reason why both fission and fusion can release energy lies in the relative stability of different elements. It turns out that elements in the middle of the periodic table of elements – such as barium and krypton (see uranium fission below) – are generally more stable than either lighter elements such as hydrogen or heavier elements such as uranium. Thus the fusion of lighter elements and fission of heavier elements are both processes which can yield more stable elemental products and this results in a release of energy.

Nuclear fission

Many large, and even some small atoms undergo nuclear fission reactions naturally. One of the isotopes of carbon – isotopes are atoms of a single element with different numbers of neutrons – called *carbon-14* behaves in this way. Carbon-14 exists at a constant concentration in natural sources of carbon. Thus living entities which constantly recycle their carbon maintain this constant concentration. However when they die, the carbon-14 is no longer renewed and it gradually decays. Measuring the residual concentration gives a good estimate of the time since the organism died. It is this property which allows archaeologists to use carbon-14 to date ancient artefacts and remains.

Other atoms can be induced to undergo fission by bombarding them with subatomic particles. One of the isotopes of uranium, the element most widely used in nuclear reactors, behaves in this manner.

Naturally occurring uranium is composed primarily of two slightly different isotopes called *uranium-235* and *uranium-238* (the numbers refer to the sum of protons and neutrons each atom contains). Most uranium is uranium-238, but 0.7% is uranium-235.

When an atom of uranium-235 is struck by a neutron it may be induced to undergo a nuclear fission reaction. The most frequent products of this reaction are an atom of krypton, an atom of barium, three more neutrons and a significant quantity of energy.

$$^{235}_{92}\mathrm{U} + \mathrm{n} = {}^{140}_{56}\mathrm{Ba} + {}^{96}_{36}\mathrm{Kr} + 3\mathrm{n} + @200\,\mathrm{meV}$$

In theory each of the three neutrons produced during this reaction could cause three more atoms of uranium-235 to split. This would lead to a rapidly accelerating reaction, called a *chain reaction*, which would release an enormous amount of energy. A chain reaction of this type forms the basis for the atomic bomb.

In fact a lump of natural uranium will not explode because the uranium-235 atoms will only react when struck by slow moving neutrons; the ones created during the fission process move too fast to cause further fission reactions to take place. They need to be slowed down first. This is crucial to the development of nuclear power.

Controlled nuclear reaction

If uranium fission is to be harnessed in a power station, the nuclear chain reaction must first be tamed. The chain reaction is explosive and dangerous. It must be curbed by both slowing the neutrons released by each fission reaction, by carrying away the energy and by controlling the neutron numbers.

As seen above, the chain reaction takes place when each fission reaction causes more than one further identical reaction. If the fission of a single uranium-235 atom causes only one identical reaction to take place, the reaction will carry on indefinitely – or at least until the supply of uranium-235 has been used up – without accelerating. But if each fission reaction leads to an average of less than one further reaction, the process will eventually die away naturally.

The operation of a nuclear reactor is based on the above idea that a nuclear chain reaction can be controlled so that the process continues indefinitely, but is never allowed to run away and become a chain reaction. A reactor in which each nuclear reaction produces one further nuclear reaction is described as critical. Once the product of each nuclear reaction is more than one further reaction, the reactor is described as supercritical.

A nuclear reactor contains uranium which has generally been enriched so that it contains more uranium-235 than it would in nature. Enrichment to about 3% is common. Using enriched uranium makes it easier to start a sustained nuclear fission reaction.

In addition to the uranium, the reactor also contains rods made of boron. Boron will absorb the neutrons generated during the nuclear reaction of uranium-235, removing them and stopping the chain reaction from proceeding. By moving the rods in and out of the reactor core, the nuclear process can be controlled.

One further crucial component is needed to make the reactor work, something to slow the fast neutrons down. The neutrons from each uranium-235 fission move too fast to stimulate a further reaction but they can be slowed by adding a material called a *moderator*. Water makes a good moderator and is used in most operating reactors. Graphite also functions well as a moderator and has been used in some reactor designs.

When a uranium fission reaction takes place the energy it releases emerges as kinetic energy. The products of the fission process carry the energy away as energy of motion, and they move extremely fast. Much of this energy is carried away by the fast neutrons. These neutrons will dissipate their energy in collision with atoms and molecules within the reactor core. In many reactors this energy is absorbed by the moderator, water. So while the neutrons are slowed, the water within the core becomes hotter. By cycling the water through the reactor core this heat can be extracted and used to generate electricity. This helps maintain the reactor in a stable condition.

Fusion

The alternative energy-yielding nuclear reaction to fission is fusion. Fusion is the process that generates energy in the sun and stars. In the sun, hydrogen atoms combine to produce helium atoms and release energy. The reaction takes place at 10–15 million°C and at enormous pressure.

The conditions in the sun cannot be recreated on earth, so here a different fusion reaction has been studied, involving two isotopes of hydrogen called *deuterium* and *tritium*. These differ in the number of neutrons their atoms contain: deuterium contains two and tritium three. Deuterium ($^{2}_{1}H$) is found naturally in small quantities in water while tritium ($^{3}_{1}H$) is made from lithium. These two will react to produce helium and energy.

$$^{2}_{1}H + ^{3}_{1}H = ^{4}_{2}He + n + @18\,meV$$

The reaction between deuterium and tritium will only take place at 100 million°C. At this temperature the atoms involved separate into nuclei and electrons, a state called a *plasma*. Since the constituents of a plasma are all charged, either positively or negatively, both can be controlled and contained using a magnetic field. This is crucial since there is no material that can withstand temperatures this severe. The most promising magnetic field shape for containing a plasma is torroidal and this has formed the basis for most fusion research. However, while fusion has been demonstrated, its commercial realisation remains a long way off.

Nuclear reactors

Nuclear reactor is the name given to the device or structure in which a controlled nuclear reaction takes place. There are a number of different designs but these have many features in common.

The core of the reactor is its heart, the place where the nuclear fuel is placed and where the nuclear reaction takes place. The fuel is most frequently formed into pellets roughly 2 cm in diameter and 1 cm long. These pellets are loaded into a fuel rod, a hollow tube of a special corrosion-resistant metal; this is frequently a zirconium alloy. Each fuel rod is 3–4 m long and a single reactor core may contain close to 50,000 such rods. Fuel rods must be replaced once the fissile uranium-235 they contain has been used up. This is a lengthy process which can take as much as 3 weeks to complete.

In between the fuel rods there are control rods, made of boron, which are used to control the nuclear reaction. These rods can be moved in and out of the core. The core will also contain a moderator to slow the neutrons released by the fission of uranium atoms. In some cases the moderator is also the coolant used to carry heat away from the core.

The outside of the core may be surrounded by a material which acts as a reflector to return some of the neutrons escaping from the core. This helps maintain a uniform power density within the core. There may also be a similar reflecting material in the centre of the core.

The coolant collects heat within the core and transfers to an external heat exchanger where it can be exploited to raise steam to drive a steam turbine. The coolant may be water (light water), deuterium (heavy water), a gas such as helium or a metal such as sodium. The core and its ancillary equipment is normally called the *'nuclear island'* of a nuclear power plant while the boiler, steam turbine and generator are called the *'conventional island'*. The coolant system will link the nuclear and conventional islands.

A nuclear power plant will contain a host of systems to ensure that the plant remains safe and can never release radioactive material into the environment. The most important of these is the containment. This is a heavy concrete and steel jacket which completely surrounds the nuclear reactor. In the event of a core failure it should be able to completely isolate the core from the surroundings and remained sealed, whatever happens within the core.

Boiling water reactor

The boiling water reactor (BWR) uses ordinary water (light water) as both its coolant and its moderator. In the BWR the water in the reactor core is permitted to boil, and the steam generated is used directly to drive a steam turbine. This steam is then condensed and recycled back to the reactor core.

This arrangement represents probably the simplest possible for a nuclear reactor because no additional steam generators are required. However the internal systems within a BWR are complex. Steam pressure and temperature are low compared to a modern coal-fired power plant and the steam turbine is generally very large. BWRs have capacities of up to 1400 MW and an efficiency of around 33%.

The BWR uses enriched uranium as its fuel. This fuel is placed into the reactor in the form of uranium oxide pellets in zirconium alloy tubes.

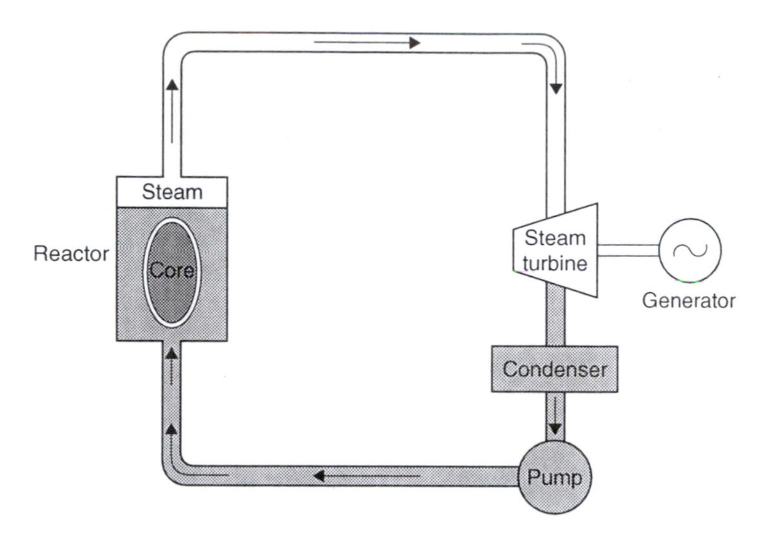

Figure 17.1 *Diagram of a BWR*

Refuelling a BWR involves removing the top of the reactor. The core itself is kept under water, the water shielding operators from radioactivity.

In common with all reactors, the fuel rods removed from a BWR reactor core are extremely radioactive and continue to produce energy for some years. They are normally kept in a carefully controlled storage pool at the plant before, in principle at least, being shipped for either reprocessing or final storage.

Pressurised water reactor

The pressurised water reactor (PWR) also uses ordinary or light water as both coolant and moderator. However in the pressurised water system the cooling water is kept under pressure so that it cannot boil.

The PWR differs in another respect from the BWR; the primary coolant does not drive the steam turbine. Instead heat from the primary water cooling system is captured in a heat exchanger and transferred to water in a secondary system. It is the water in this second system which is allowed to boil and generate steam to drive the turbine.

The use of a second water cycle introduces energy losses which make the PWR less efficient at converting the energy from the nuclear reaction into electricity. However the arrangement has other advantages regarding fuel utilisation and power density, making it competitive with the BWR.

The PWR uses enriched uranium fuel with a slightly higher enrichment level than in a BWR. This is responsible for a higher power density within the reactor core. As with the BWR, the fuel is introduced into the core in the

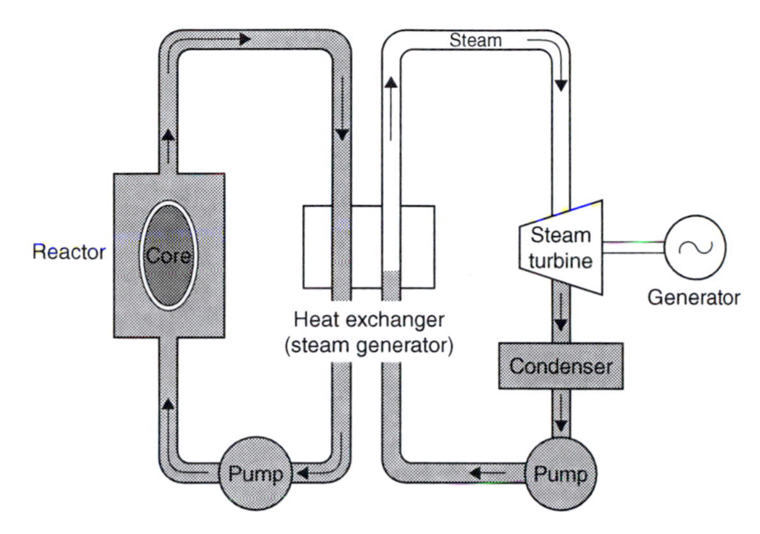

Figure 17.2 *Diagram of a PWR*

form of uranium oxide pellets. A typical PWR has a generating capacity of 1000 MW. The efficiency is around 33%.

Canadian deuterium uranium reactor

The Canadian deuterium uranium (CANDU) reactor was developed in Canada with the strategic aim of enabling nuclear power to be exploited without the need for imported enriched uranium. Uranium enrichment is an expensive and highly technical process. If it can be avoided, countries such as Canada with natural uranium reserves can more easily exploit their indigenous reserves to generate energy. This has made the CANDU reactor, which uses unenriched uranium, attractive outside Canada too.

The CANDU reactor uses, as its moderator and coolant, a type of water called *heavy water*. Heavy water is a form of water in which the two normal hydrogen atoms have been replaced with two of the isotopic form, deuterium. Each deuterium atom weighs twice as much as a normal hydrogen atom, hence the name heavy water. Heavy water occurs in small quantities in natural water.

Heavy water is much more expensive than light water but it has the advantage that it absorbs fewer neutrons. As a consequence, it is possible to sustain a nuclear reaction without the need to enrich the uranium fuel. The CANDU reactor has the additional advantage that it can be refuelled without the need to shut it down; in fact this is necessary with natural uranium fuel to keep the plant going. Avoiding lengthy refuelling shutdowns provides better operational performance.

The CANDU fuel is loaded in the form of uranium oxide pellets housed in zirconium alloy rods. Fuel replacement involves pushing a new rod into a pressure tube which passes through the vessel containing the heavy water (called a *calandria*) and forcing the old tube out of the other end.

The heavy water coolant in the CANDU reactor is maintained under pressure so that it cannot boil. Heat is transferred to a light water system in a steam generator and the secondary system drives a steam turbine in much the same way as a PWR. Efficiency is similar too.

Advanced gas-cooled reactor

The advanced gas-cooled reactor (AGR) is a specifically UK breed of reactor that was developed from the design for the very first nuclear reactor to generate electricity for commercial use, a reactor built at Calder Hall in Cumbria, UK. The AGR employs a graphite moderator and uses carbon dioxide as its coolant.

The graphite moderator in an AGR is pierced by a series of channels into which fuel rods are placed. The rods, clad in zirconium alloy, contain uranium in the form of uranium oxide, enriched with 2% uranium-235. Carbon dioxide floods the core. This carbon dioxide carries the heat generated by fission in the reactor to a heat exchanger where it is used to generate steam to drive a turbine.

Several AGRs have been built in the UK but these have been found to be more costly to operate than was initially anticipated and no further units of this design are planned. Instead, the last nuclear power plant built in the UK employed a US PWR design.

High-temperature gas-cooled reactor

The high-temperature gas-cooled reactor (HTGR) is similar in concept to the AGR. It uses uranium fuel, a graphite moderator and a gas as coolant. In this case, however, the gas is helium.

Several attempts have been made to build reactors of this type but none has so far entered commercial service. Early development work was carried out in the USA. The US design utilised fuel elements in the shape of interlocking hexagonal prisms of graphite containing the fissile material. HTGR fuel is often much more highly enriched than the fuel in a water-cooled reactor, with up to 8% uranium-235. The arrays of hexagonal graphite prisms contain shafts for control rods and passages for the helium to pass through and carry away the heat generated by fission.

Another design, developed in Germany, uses uranium oxide fuel which is sealed inside a graphite shell to form a billiard ball-sized fuel element called a *pebble*. This gives the reactor its name, the pebble-bed reactor. Development of this in Germany was eventually abandoned but the idea was taken up during the 1990s by the South African utility Eskom which is still developing the design. Japan and China have experimental programmes too.

The advantage of the HTGR is that both the moderator, graphite, and the coolant, helium, can operate at high temperature without reacting or deteriorating. A typical HTGR will operate at a pressure of 100 atm and at a higher temperature than a water-cooled reactor. This enables better thermodynamic operation to be achieved. The reactor is designed so that in the event of a coolant failure it will be able to withstand the rise in internal temperature without failing.

The HTGR can use a dual cycle system in which the helium coolant passes through a heat exchanger where the heat is transferred to water and steam is generated to drive a steam turbine. This arrangement is around 38% efficient. However a more advanced system uses the helium directly to drive a gas turbine. This arrangement is sometimes called a *gas turbine modular helium reactor* (GT-MHR). In theory the GT-MHR can achieve an energy conversion efficiency of 48%.

One of the attractions of the HTGR is that it can be built in relatively small unit sizes. Modules can have generating capacities of between 100 MW and 200 MW, making it attractive for a wider variety of applications. The modular form of most designs also makes it easy to expand a plant by adding new modules. However no reactors of this design have yet entered commercial service.

Breeder (fast) reactors

The breeder reactor uses, not uranium, but plutonium as fuel. This decays or splits in a similar way to uranium-235, producing fast neutrons. However whereas in the conventional uranium-235 reactor the fast neutrons must be slowed with a moderator to enable further nuclear fusion reactions to take place, the fast reactor utilises the fast neutrons so no moderator is required.

The unique feature of the breeder reactor is that the reactor core contains, in addition to plutonium, some uranium-238. Uranium-238, the more common isotope of uranium, can capture the fast neutrons, becoming converted into plutonium in the process.

By careful design it is possible to make a breeder reactor actually produce (or breed) more plutonium than it burns, hence the name. Breeder reactors use liquid sodium as the coolant because the material does not slow down fast neutrons. However the use of this coolant can create severe technical problems.

Several countries have developed breeder reactors but none has entered full commercial service. The coolant, liquid sodium, has proved the Achilles heel in at least two projects, one in France and one in Japan. It is not clear that the nuclear fast reactor has any future for power generation.

Advanced reactor designs

There are a number of advanced reactor designs being developed across the world. These include the HTGR reactors discussed above as well as development of the various water-cooled designs. The latter are mainly aimed at improving safety and reducing the cost of construction. Passive safety features which operate in a failsafe fashion if any part of the reactor system fails are being pursued in many designs and modular construction is seen as a key to reducing overall construction cost and time.

Nuclear fusion

The development of nuclear fusion has a history stretching back more than 50 years yet a commercial power plant based on the technology could still be 50 years away. The fusion reaction requires a temperature of 100 million°C. At this temperature all matter exist in a state called a plasma. The plasma must be controlled and contained by a magnetic field. There are no materials capable of resisting 100 million°C without becoming plasmas themselves.

Research into nuclear fusion has focussed on a torroidal magnetic containment for the fusion reaction, the most successful of which has been a design called a *Tokamak*. Tokamak's have been tested in experimental fusion reactors but no fusion reactor has yet been able to generate more energy than has been supplied to it. That is, the aim of an international project.

The next stage in fusion research and development is a project called *International Thermonuclear Experimental Reactor* (ITER), a project involving a large group of nations including the USA, Russia, Japan, the EU and China. The aim of ITER is to build a 500 MW fusion reactor to prove the concept. This is likely to cost around €4.5 billion but should be finished by the end of the first decade of the twenty-first century. If it is successful it could pave the way for the first generation of fusion power stations towards the middle of the twenty-first century.

Environmental considerations

The use of nuclear power raises important environmental questions. It is an apparent failure to tackle these satisfactorily that has led to much of the popular disapprobation that the nuclear industry attracts. There are two adjuncts to nuclear generation that cause the greatest concern, nuclear weapons and nuclear waste.

While the nuclear industry would claim that the civilian use of nuclear power is a separate issue to that of atomic weapons, the situation is not that clear cut. Nuclear reactors are the source of the plutonium which is a primary constituent of modern nuclear weapons. Plutonium creation depends on the reactor design; a breeder reactor can produce large quantities while a PWR produces very little. Nevertheless all reactors produce waste that contains dangerous fissile material. This is a subject of international concern.

The danger is widely recognised. Part of the role of the International Atomic Energy Agency is to monitor nuclear reactors and track their inventories of nuclear material to ensure than none is being sidetracked into nuclear weapons construction. Unfortunately, this system can never be foolproof. It seems that only if all nations can be persuaded to abandon nuclear weapons can this danger, or at least the popular fear of it, be removed. At the beginning of the twenty-first century such an agreement looks highly improbable.

The problem is political in nature. Nevertheless it carries a stigma from which the industry can never escape. The prospect of a nuclear war terrifies most people. Unfortunately for the nuclear power industry, some of the after effects of nuclear explosion can also be produced by a major civilian nuclear accident.

The contents of a nuclear reactor core includes significant quantities of extremely radioactive nuclei. If these were released during a nuclear accident they would almost inevitably find their way into humans and animals via the atmosphere or through the food chain.

Large doses of radioactivity or exposure to large quantities of radioactive material kills relatively swiftly. Smaller quantities of radioactive material are lethal too, but over longer time scales. The most insidious effect is the genesis of a wide variety of cancers, many of which may not become apparent for 20 years or more. Other effects include genetic mutation which can lead to birth defects.

The prospect of an accident leading to a major release of radionucleides has created a great deal of apprehension about nuclear power. The industry has gone to extreme lengths to tackle this apprehension by building ever more sophisticated safety features into their power plants. Unfortunately the accidents at Three Mile Island in the USA and Chernobyl in the Ukraine remain potent symbols of the danger.

This danger has been magnified by the rise of international terrorism. The threat now exists that a terrorist organisation might create a nuclear power plant accident, or by exploiting contraband radioactive waste or fissile material, cause widespread nuclear contamination.

So far a nuclear incident of catastrophic proportions has been avoided. Smaller incidents have not, and low-level releases of radioactive material have taken place. The effects of low levels of radioactivity have proved difficult to quantify. Safe exposure levels are used by industry and regulators but these have been widely disputed. Only the elimination of radioactive releases from civilian power stations is likely to satisfy a large sector of the public.

Radioactive waste

As the uranium fuel within a nuclear reactor undergoes fission, it generates a cocktail of radioactive atoms within the fuel pellets. Eventually the fissile uranium becomes of too low a concentration to sustain a nuclear reaction. At this point the fuel rod will be removed from the reactor. It must now be disposed of in a safe manner. Yet after more than 50 years, no safe method of disposal has been developed.

Radioactive waste disposal has become one of the key environmental battlegrounds over which the future of nuclear power has been fought. Environmentalists argue that no system of waste disposal can be absolutely safe, either now and in the future. And since some radionucleides will remain a danger for thousands of years, the future is an important consideration.

Governments and the nuclear industry have tried to find acceptable solutions. But in countries where popular opinion is taken into consideration, no mutually acceptable solution has been found. As a result, most spent fuel has been stored in the nuclear power plants where it was produced. This is now causing its own problems as storage ponds designed to store a few years' waste become filled, or overflowing.

One avenue that has been explored is the reprocessing of spent fuel to remove the active ingredients. Some of the recovered material can be recycled as fuel. The remainder must be stored safely until it has become inactive. But reprocessing has proved expensive and can exacerbate the problem of disposal rather than assisting it. As a result it appears publicly unacceptable.

The primary alternative is to bury waste deep underground in a manner that will prevent it ever being released. This requires both a means to encapsulate the waste and a place to store the waste once encapsulated. Encapsulation techniques include sealing the waste in a glass-like matrix.

Finding a site for such encapsulated waste has proved problematical. An underground site must be in stable rock formation is a region not subject to seismic disturbance. Sites in the USA and Europe have been studied but none has yet been accepted. Even if site approval is achieved, there appears little prospect of any nuclear waste repository being built until well into the second decade of the twenty-first century.

Other solutions have been proposed for nuclear waste disposal. One involves loading the fuel into a rocket and shooting it into the sun. Another utilises particle accelerators to destroy the radioactive material generated during fission.

Environmentalists argue that the problem of nuclear waste is insoluble and represents an ever-growing burden on future generations. The industry disputes this but in the absence of a persuasive solution its arguments lack weight. Unless a solution is found, the industry will continue to suffer.

Waste categories

Spent nuclear fuel and reprocessing plant waste represent the most dangerous of radioactive wastes but there are other types too. In the USA these first two types of waste are categorised as high-level waste[8] while reminder of the waste from nuclear power plant operations is classified as low-level waste. There is also a category called *transuranic waste* which is waste containing traces of elements with atomic numbers greater than that of uranium (92). Low-level wastes are further subdivided into classes depending on the amount of radioactivity per unit volume they contain.

In the UK there are three categories of waste, high level, intermediate level and low level. High level includes spent fuel and reprocessing plant waste, intermediate level is mainly the metal cases from fuel rods and low-level waste constitutes the remainder. Normally both high- and intermediate-level waste require some form of screening to protect workers while low-level waste can be handled without a protective radioactive screen.

High-level wastes are expected to remain radioactive for thousands of years. It is these wastes which cause the greatest concern and for which some storage or disposal solution is most urgently required. But these wastes form a very small part of the nuclear waste generated by the industry. Most is low-level waste. Even so it too must be disposed of safely. Low-level waste can arise from many sources. Anything within a nuclear power plant that has even the smallest exposure to any radioactive material must be considered contaminated. One of the greatest sources of such waste is the fabric of a nuclear power plant itself.

Decommissioning

A nuclear power plant will eventually reach the end of its life and when it does it must be decommissioned. At this stage the final, and perhaps largest nuclear waste problem arises.

After 30 or more years[9] of generating power from nuclear fission, most of the components of the plant have become contaminated and must be treated as radioactive waste. This presents a problem that is enormous in scale and costly in both manpower and financial terms.

The cleanest solution is to completely dismantle the plant and dispose of the radioactive debris safely. This is also the most expensive option. A half-way solution is to remove the most radioactive components and then seal up the plant for from 20 to 50 years, allowing the low-level waste to decay, before tackling the rest. A third solution is to seal the plant up with everything inside and leave it, entombed, for hundreds of years. This has been the fate of the Chernobyl plant.

Decommissioning is a costly process. Regulations in many countries now require that a nuclear generating company put by sufficient funds to pay for decommissioning of its plants. In the USA, studies suggest that the cost of decommissioning a nuclear plant will be around $370 million. The total US bill for decommissioning its nuclear plants is expected to reach $40 billion. When building a new nuclear plant, the cost of decommissioning must, therefore, be taken into account.

Financial risks associated with investing in nuclear power

Nuclear power generation technology is a mature technology and is well understood. Construction of a new nuclear plant based on established technology should present no significant technical risk.

Where innovations are made to nuclear power plant designs, these are usually evolutionary in nature, based clearly on existing technology. The nuclear industry has found this approach to be essential because of the difficulty in obtaining authorisation to build novel nuclear plants. Technological risk should, therefore, remain low even where changes to plant design have been instigated.

The most significant nuclear risks lie elsewhere. Nuclear power is capital intensive. The cost of the plant is much higher than that of a fossil-fuelled power plant but the cost of the fuel is much lower. This makes nuclear plant construction extremely sensitive to schedule overruns.

In the USA in the later stages of its development, nuclear power plants were taking up to 10 years to build. Over this period interest rates can change dramatically, fuel costs can change, and perhaps most significant of all, regulations can change.

The introduction of new regulations affecting the construction of nuclear power stations can easily affect the construction schedule by years. Then interest payments escalate. It was the conspiracy of just such factors in the USA which pushed several utilities with nuclear construction programmes close to bankruptcy.

The route around such problems is with standardised designs which can be authorised rapidly and modular construction techniques to ensure rapid construction schedules. If the construction schedule can be kept short then the risk becomes significantly lower. A 1300 MW reactor which was commissioned in Japan in 1996 took little over 4 years to complete. Construction periods of 4 years or less are essential in the future.

There remains the risk of a nuclear accident. There may even be liability in the event of a terrorist incident. Any nuclear power company must attempt to indemnify itself against this possibility. The claims that might be made as a result of a significant release of radioactive material are incalculable but undoubtedly gargantuan.

The cost of nuclear power

Nuclear power is capital intensive and costs have escalated since the early days of its development. This is partly a result of higher material costs and high interest rates but is also a result of the need to use specialised construction materials and techniques to ensure plant safety. In the USA, in the early 1970s, nuclear plants were being built for units costs of $150–300/kW. By the late 1980s, the figures were $1000–3000/kW.

The Taiwan Power Company carried out a study, published in 1991, which examined the cost of building a fourth nuclear power plant in Taiwan. The study found that the cost for the two-unit plant would be US$6.3 billion, a unit cost of around $3150/kWh. The estimate was based on completion dates of 2001 and 2002 for the two units. Orders were actually placed in 1996, with construction now scheduled for completion in 2004 and 2005.

Nuclear construction costs do not take into account decommissioning. This can cost from 9% to 15% of the initial capital cost of the plant. However nuclear proponents argue that when this is discounted it adds only a few percent to the investment cost.

The fuel costs for nuclear power are much lower than for fossil-fuel-fired plants, even when the cost of reprocessing or disposal of the spent fuel are taken into account. Thus, levelised costs of electricity provide a more meaningful picture of the economics of nuclear power generation.

Table 17.2 gives figures from the 1991 Taiwan Power Company study which shows levelised costs of generation of power from different sources, based on a plant with a 25-year lifetime which starts operating in 2000. Nuclear power, at T$2.703/kW, is cheaper than the other sources of power cited. Actual unit generation costs from existing plants in 1997 are provided for comparison. Again nuclear power is the cheapest source, closely followed by coal and hydro.

Table 17.2 *Cost of power generation in Taiwan*

	25-year levelised cost (T$)	*Unit generation cost in 1997 (T$)*
Nuclear	2.703	0.89
Coal fired	3.023	1.00
Oil fired	4.136	1.39
LNG fired	4.462	2.04
Hydro	–	1.03

Note: Levelised costs are based on a 25-year lifetime from 2000 to 2025.
Source: Taiwan Power Company.

Taiwan has to import all its fuel so costs for fossil-fuel-fired generation are bound to be high. Where cheap sources of fossil fuel are available locally, the situation will be different. Australia, for example, estimates that coal-fired power generated a pithead plants is cheaper than nuclear power.

A 1997 European study compared the cost of nuclear-, coal- and gas-based power plants for base-load generation. For a plant to be commissioned in 2005, nuclear power was cheaper than all but the lowest-priced gas-fired scenario, based on a discount rate of 5%. When the discount rate was put up to 10%, nuclear power was virtually the most expensive option. Other studies have confirmed this assessment.

Coal is generally the source of new generating capacity with which nuclear investment is compared. But the cost of coal, and therefore the cost of coal-fired electricity, depends heavily on transportation costs. These can account for as much as 50% of the fuel cost. Given this sensitivity, the local availability of coal will be a strong determinant of the economic viability of nuclear power. Gas-fired base-load generation in combined cycle power plants is also cheap but similarly sensitive to fuel prices.

While the cost of new nuclear generating capacity might be prohibitive in some parts of the world – but acceptable in others – the cost of power from existing nuclear power plants is often extremely competitive. This is true even where coal and gas are readily available. Thus the Nuclear Energy Institute claims that 2002 was the fourth year for which nuclear-generated electricity was the cheapest in the USA, undercutting power from coal-, oil- and gas-fired power plants. (Hydropower from old plants may well be cheaper still, see Chapter 8.) In support of this, a number of companies are now making a successful business of running US nuclear power stations sold by utilities when the US industry was deregulated. In France too, nuclear power is on average the cheapest source of electricity.

End notes

1 US Department of Energy.
2 There were rumours in late 2004 that Russia was starting work on a breeder reactor at Beloyarsk.
3 World Energy Council, Survey of Energy Resources, 2001.
4 International Atomic Energy Authority.
5 World Energy Council, Survey of Energy Resources, 2001.
6 Refer *supra* note 5.
7 Nuclear Energy Institute.
8 The US Department of Energy does not classify spent fuel as waste but the Nuclear Regulatory Commission does.
9 US nuclear plants are now winning operating license extensions which will allow them to operate for up to 60 years.

Index